THE TRIANGLE FIRE

THE TRIANGLE FIRE

BY LEON STEIN

INTRODUCTION BY

WILLIAM GREIDER

ILR PRESS

An imprint of

CORNELL UNIVERSITY PRESS

Ithaca and London

First published 1962 by J. B. Lippincott and issued in paperback 1985
by Carroll & Graf Publishers/Quicksilver Books
First printing, ILR Press/Cornell Paperbacks, 2001

ISBN-13: 978-0-8014-8714-9 (pbk.: alk. paper)
ISBN-10: 0-8014-8714-5 (pbk.: alk. paper)

Printed in the United States of America

Librarians: A CIP catalog record for this book is available from the Library of Congress.

Cornell University Press strives to use environmentally responsible suppliers and materials to the fullest extent possible in the publishing of its books. Such materials include vegetable-based, low-VOC inks and acid-free papers that are recycled, totally chlorine-free, or partly composed of nonwood fibers. For further information, visit our website at www.cornellpress.cornell.edu.

Paperback printing 10 9 8 7 6 5

INTRODUCTION

"Who Will Protect the Working Girl?"

WILLIAM GREIDER

You are about to read a harrowing story of human struggle from America's past. With suspenseful intimacy, this book recounts a great tragedy that occurred nearly one hundred years ago yet lingers in the national memory—the deaths of 146 young women in a disastrous garment-factory fire in New York City. The Triangle fire in 1911 still gets a mention in annual editions of the U.S. almanac because it is the worst industrial factory fire in the history of American capitalism.

Leon Stein tells this story in brilliant cinematic fashion, with quick jump-cuts in the action, moving from one scene to the next without the interruption of commentary. You see the events directly, as if you were standing on the street outside the burning factory. The experience is a little like watching a well-made horror film—you want to know what happens next but are afraid to find out. Only this story is terribly real and involves actual human beings who suffer and die. You will feel a quickening anxiety as the story unfolds, then anger as you begin to grasp what killed those working girls.

As we learn more about these young women, we see that they were victims of extraordinary callousness and dereliction on the part of important and powerful people. The factory owners, of course, but also the city's business community, the elected officials of New York, and the various bureaucracies responsible for enforcing public safety and health standards—all could be blamed, in one way or another, for the unseen dangers women faced at the Triangle Shirtwaist Company, where they manufactured the shirtwaist blouses that were the reigning fashion in the early twentieth century. But, in a larger sense, New Yorkers as a whole were also culpable—as was the nation—because for many years people had generally ignored the cries of protest from these "girls" and from the millions of other industrial workers laboring in similarly deplorable circumstances.

Many of these workers were immigrants, typically trapped in miserable and dangerous working conditions, struggling for a meager livelihood in their new country. Americans, meanwhile, kept their distance. They viewed the newcomers as dirty and dangerous, peculiar foreigners who spoke strange languages and lived in congestions of squalid housing (then and now often called ghettoes). The jobs the immigrants performed were vital to America's booming manufacturing, yet most Americans barely knew these people existed. Those who did were often wary and contemptuous.

Does this sound familiar? The very same conditions have returned in our own time, both within the United States and around the world. Sweatshops are back in Manhattan, operating illegally with new immigrant workers and within blocks of the site of the old Triangle factory. Sweatshops are operating across the nation in sectors from food processing to apparel. At the same time, the young "working girls" of Asia, Mexico, Central America, and elsewhere are now manufacturing the goods we buy, from expensive name-brand sports shoes to designer jeans to toys and high-end electronics. They are often paid pitiful wages and subjected to degrading, sometimes dangerous, working conditions. Aside from the modern technologies, the main differences in American sweatshops today are the names and faces of the new immigrants. A century ago, they were overwhelmingly European—Italian, Irish, Polish, or, like the Triangle workers, Jews who had fled Russia or the Ukraine for the promise of freedom in American life. Today, the immigrant workers are mainly from Mexico and other Latin countries, or the rim of Asia, from Cambodia and Vietnam to Korea and China. Joined by a multitude of U.S. citizens, especially Afro-Americans, these foreign-born workers do the dirty, low-wage jobs necessary to the economic system. And they are, likewise, ignored or despised.

Why is this happening again? Weren't sweatshop conditions banned by law generations ago? What is the economic necessity of repeating old inhumanities in these modern times of great abundance? You should be asking these questions, among others, while reading this book. The grim fact is that young workers in Asia are once again dying needlessly in burning factories and for approximately the same reasons the young women died in the Triangle fire. Most of these workers are young women. Some are mere children. And American consumers have hardly noticed.

This introduction provides a context for understanding what happened at the Triangle disaster and why it became a pivotal event in American politics. But my more pressing point is that this long-ago story is savagely relevant today. We may avert our gaze and ignore the realities, just as many people did in the past, but the injustices are not going to go away on their own. Sooner or later, events will compel us to face them, just as Americans had to when they were shocked and embarrassed by the Triangle fire.

While we focus our sympathies on the victims in this story, we must not reduce them to inert, hapless objects of pity. Certainly, their lives were hard, but the young women who worked at the Triangle Shirtwaist Company were like other young adults new to the big city, joyfully soaking up its excitement and absorbed in their own ambitions. A contemporary artist who sketched young workers on the streets of Manhattan's Lower East Side described them as "a bevy of boisterous girls with plenty of energy left after a hard day's work." Their wages were meager, but the "working girls," as they were called in those days, typically brought home income that was essential to their family's survival. Despite the hardships, it was a promising time for them, and they knew it.

This was the era of America's great industrial advancement, 1900 to 1920, when new manufacturing giants in steel, automobiles, and electrical generation rose to preeminence, producing marvelous, new conveniences like cars, telephones, refrigerators, and indoor plumbing. It was also a time of intense conflict on social and political fronts, because the rising tides that would lead eventually to modern prosperity also generated bitter contests between the haves and have-nots, between workers and companies. At the same time consumers were enjoying dazzling comforts, reformers were campaigning against child labor and unsafe factories, and communities were defending themselves against slums and slumlords and other damaging social consequences of industrialization.

The young women employed at Triangle and other garment factories concentrated in New York City did not need to be told they were being exploited. They knew it from their wages and working conditions. A couple of years before the tragedy occurred, these same young immigrant women mobilized themselves and staged an audacious demonstration of their grievances—a general, industry-wide strike against their employers. Their protest became known as "The Uprising of the Twenty Thousand," and, at the time, it was the largest strike ever organized by working women anywhere in the world. Consider for a moment the courage that required of them.

Their situation demanded collective action. The supply of willing workers was so abundant that whenever any individual worker made complaints, she was easily fired and replaced. Their general strike in 1909 was a sensational event—actually an intriguing novelty in an era when labor unions were weak or nonexistent for men in industrial jobs, even more so for women workers who had not even won the right to vote. And the collective withdrawal of their labor won some partial victories. Many employers, especially the smaller operators, were compelled to agree to better terms of employment or to accept a union to represent the workers. Many other employers, including the Triangle Shirtwaist Company, did not yield.

Their uprising occurred at the birth of the Ladies International Garment Workers Union, which survives today incorporated in UNITE, the Union of Needletrades, Industrial, and Textile Employees (the author of this book was for many years editor of *Justice*, ILGWU's magazine). The efforts of industrial workers to gain a voice and a fair share of the returns from their labor produced many similar confrontations, stretching across decades. By its nature, it was a long and tortuous struggle and indeed continues in our time. UNITE and a number of other unions are now waging a difficult fight to organize new immigrants wherever they work, from dangerous jobs in meatpacking houses in the Midwest to the modern high-rise office buildings, from New York to Los Angeles, where the janitors who mop and clean at night also need collective action to win decent wages.

A century ago, against this background of social protest and labor's demands, the Triangle fire became a galvanizing moment in American history. It provoked nationwide shock and anger but also a humiliating sense of guilt among many citizens who had previously ignored labor's grievances. "Who will protect the working girl?" became the popular rallying demand, and it spawned a great social movement, stirred political action on many fronts, and eventually led to genuine reforms. The labor movement itself was energized and became more aggressive. Middle-class reformers were emboldened in their demands for reform laws on health, safety, and the scandal of children working in the mills and mines. Along with organized labor, reformers aroused public concern across a broad spectrum of Americans. And some leading politicians bravely took up the cause, enacting ordinances and statutes in city councils and state legislatures, especially New York's.

Even so, the struggle for reform was difficult and slow, with many political setbacks. New York and a few other progressive states enacted landmark laws establishing new social standards and regulatory rules for industrial enterprise (some of which were initially nullified by the Supreme Court as improper intrusions on private business). But it was twenty-five years before the federal government acted. The rights of workers to organize and other great social reforms were finally consecrated in national law by President Franklin D. Roosevelt in the 1930s, the era known as the New Deal. FDR's secretary of labor, a reformer from New York named Frances Perkins (the first woman cabinet officer in U.S. history), remarked at the time: "The Triangle fire was the first day of the New Deal."

New York City was terribly divided in that earlier period of industrial conflict—divided between the well-to-do and the impoverished, between labor and industrialists. But the city was also bitterly divided by its multiplicity of hostile ethnic groups—immigrants against immigrants. The Triangle catastrophe pulled these different people closer together and helped them recognize their common humanity. A friend from New York once explained to me how that happened. "The Irish cops were picking up the bodies of the Jewish girls," he said, "and that changed New York politics forever."

Fast-forward to May 10, 1993. A huge toy factory with three thousand workers is destroyed by fire in an industrial zone on the outskirts of Bangkok, Thailand—188 workers killed and 469 severely injured. The actual toll was higher because the four-story building swiftly collapsed, and many bodies were incinerated in the intense flames, some never to be found. All but fourteen of the dead were women, most quite young and some as young as thirteen years old. The mammoth factory had been poorly designed and cheaply built, with inadequate stairways and structure, and no fire-prevention equipment whatsoever. Managers at the Kadar Industrial Toy Company had kept exit doors locked to prevent workers from pilfering the stuffed toys and plastic dolls that they manufactured. Hundreds of workers were trapped on the upper floors, forced to jump to their deaths or to survive with broken backs and limbs. Nine months after the fire, I visited Thailand while working on a book about the global economy and found that physical evidence of the disaster was gone—the factory site had been scraped clean, the company's toy production moved to a quickly-built factory in a remote rural location.

Introduction

The toy factory fire in Thailand was the worst in the history of capitalism—surpassing even the Triangle fire—yet the rest of the world barely noticed. It was reported on page twenty-five of the *Washington Post*. The *Wall Street Journal* followed a day late with a brief account on page eleven. The *New York Times* also put the story inside, though it did print a chilling photo on the front page—rows of small, shrouded bodies lying on bamboo pallets, with dazed rescue workers standing amid the corpses. Catastrophes in distant foreign countries are simply not considered big news for Americans unless hundreds of thousands have died.

What the news stories failed to mention was that these Thai workers assembled toys for American children—the most popular items such as Bart Simpson, the Muppets, Big Bird and other Sesame Street characters, Playskool "Water Pets," Bugs Bunny, and Santa Claus dolls. The factory, jointly owned by Thailand's leading conglomerate and a major Taiwanese manufacturer, supplied the products sold by the biggest names in American toys—Fisher-Price, Hasbro, Tyco, Kenner, Toys "R" Us, J. C. Penny, and others. The U.S. companies easily distanced themselves from any responsibility. Their customers remained largely ignorant of the tragic callousness that lurked behind the toys they bought for their children.

In my reporting, I talked to some of the workers who had witnessed the event and had searched among the bodies for friends and relatives, attending to the maimed survivors. I also interviewed Thai labor leaders and sympathetic university professors who had helped the workers organize protests afterwards, demanding just compensation for the victims' children and families. These people were under the impression that a worldwide boycott of Kadar's products was occurring on their behalf. Public opinion in America, they assumed, was demanding reforms that would compel global companies to adhere to elementary standards for safe factories by installing modern fire-prevention equipment in well-constructed buildings. I had to inform them there was no boycott. I did not have the heart to tell them that most Americans had never even heard of their tragedy.

The details of how the Thai women at the Kadar Industrial Toy Company died are hauntingly similar to the Triangle fire deaths eight decades earlier in New York City. Let us listen to the testament from just one of the survivors, a young woman named Lampan Taptim:

> There was the sound of yelling about a fire. I tried to leave the section, but my supervisor told me to get back to work. My sister who worked on the fourth floor with me pulled me away and insisted we try to get out. We tried to go down the stairs and got to the second floor; we found the stairs had already caved in. There was a lot of yelling and confusion. In desperation, I went back up to the windows and went back and forth, looking down below. The smoke was thick and I picked the best place to jump in a pile of boxes. My sister jumped, too. She died.

Similar tragedies, large and small, are now commonplace in many developing countries where globalization has located low-wage factories that produce goods to export to wealthier nations like the United States. Six months after Kadar, eighty-four women died and dozens were severely burned in a similar toy factory fire in the burgeoning industrial zone in Shenzhen, China. Earlier, in 1991, a raincoat factory in Dongguan had burned, killing more than eighty workers. Some sixty women died at a textile factory in Fuzhou Province. There have been other instances. "Why must these tragedies repeat themselves again and again?" the *People's Daily* in Beijing asked. The official *Economic Daily* blamed the greed and immorality of foreign investors.

Recently, I conducted a quick survey to see if these scandalous conditions had been eliminated, either by local reformers and governments or by multinational corporations insisting that their subcontractors and suppliers clean up their act. Not much has changed. In 1999, the Zhimao Electronics fire in Shenzhen left twenty-four dead and forty injured. The *China Labor Bulletin* described it as "a copycat of at least six similar fires in South China." Nine people, including five children, died in a furniture factory in

Nanyang. In Bangladesh, the largest apparel factory in Dhaka burned to the ground on a Sunday night (evidently without casualties), and firefighters reported the absence of any fire-prevention system. One recurring element in many of these cases is that the workers—usually young women imported from impoverished rural villages—are often housed in factory dormitories with security guards at the doors. When fire breaks out, the workers either cannot escape or suffocate from the smoke and toxic fumes.

There are many more examples of such abuses in the global production system, many other dimensions of the injustice visited upon the people who are workers. Like the young women in Bangkok in 1993 or in New York City in 1911, the victims lack the status and influence to defend themselves. Who will protect the working girl? Before we get too angry at the laxity of foreign governments, we should recall that similar gruesome episodes, albeit less deadly than the Kadar or Triangle tragedies, have surfaced again in the United States. In 1991, twenty-five people died when a chicken-processing plant caught fire in Hamlet, North Carolina. The exit doors were locked. In El Monte, California, inspectors discovered a garment factory where seventy-two Thai immigrants were being held in virtual peonage, working eighteen hours a day in subhuman conditions. The U.S. Government Accounting Office surveyed the nation in the mid-1990s and reported: "In general, the description of today's sweatshops differs little from that at the turn of the century."

The modern scandal, however, is complicated in ways that make it difficult for people to grasp. First, the abusive conditions are now scattered around the world, often in obscure places that are not visible to the public. The outrage is localized and rarely, if ever, reported by the U.S. press. Second, regulatory laws already exist and not just in the United States. Most developing countries, including China and Thailand, have sound laws that require safe working conditions and even guarantee the workers' right to organize. These laws are widely unenforced and regularly evaded by businesses, including the subcontractors who supply major American multinationals. In the embarrassment following the toy factory fire, the Thai minister of industry ordered special inspections of 244 large factories in the Bangkok region and found that sixty percent of them had violations similar to Kadar's.

Why does the government tolerate this? Because, even with good intentions, the laws are subverted by fierce competition among poor countries, all of whom are desperate for foreign investment and new factories that promise jobs and growth. A country like Thailand, though developing robustly for many years, is caught in the middle. If it conscientiously asserts the higher legal standards, the factories will indeed pick up and leave for China or some other very poor nation that is prepared to tolerate abuses of its own citizens in the name of economic development. Economic progress is like a treadmill that keeps workers running in place while companies profit. The Thai minister of industry candidly admitted the dilemma globalization poses for government law enforcement: "If we punish them, who will want to invest here?"

His frustrated question is not very different from what New York City officials told themselves a century ago when they contemplated the dangerous conditions in proliferating sweatshops. If we come down too hard on these factory owners, they will simply close down and move somewhere else in America—and then where will we find jobs for all of these poor immigrants? Political corruption, of course, also accompanies the retreat from principle, then and now. It is fair to say, I think, that contemporary politicians in America (and indeed many ordinary American citizens) have accepted the same passive response to economic injustices. Important people who have the power to correct the abuses are once again looking the other way.

The trouble is, the passivity of government and the public simply leads further down a low road. More injustices appear, and they, too, must be tolerated in the name of commerce. As history demonstrates, this human degradation will continue, here and abroad, until people rouse themselves to stop it. In a global system, poorer nations (even China if it wished) cannot halt exploitation on their own. The factories *do* leave one nation for another with lower standards, and there is an endless food chain of very poor countries waiting for jobs on any terms. The process will not change until international laws build a floor of reasonable standards for business performance.

This is why the global system needs new and reformed trade and investment agreements that penalize both the multinational companies and the trading nations that choose inhumane production methods to gain a competitive advantage over rivals. The economic incentives to take the low road also explain why protecting labor rights is so important for achieving genuine progress. If workers have the freedom to speak for themselves—to organize a collective voice and demand respect and a living wage—they can become a bulwark not only for advancing the society's values and self-respect but also for raising its general standard of living. As we know from history, none of these achievements comes easily or without great effort from conscientious citizens.

In my travels, I met some of the young workers in the globalized production system, usually away from the factories in places where they could talk more honestly about their hopes and fears, ambitions and disappointments. From Indonesia and Malaysia to China and Mexico and the other countries, I always came away enthralled by their innocent struggles but also in awe of their elemental bravery. The young women and men, it is true, are often bewildered by their new circumstances as industrial workers in the complex global system. They are nevertheless trying to understand their situation. Of course, they are glad to have jobs with wage incomes, since most have migrated from poverty in rural backwaters. And, like the girls at Triangle, they are excited by their new surroundings in or near booming cities. They, too, faithfully send home money to help their impoverished families.

Yet these young workers are also rebelling against the scant pay they receive and the abusive conditions in the factories. Many of the workers I interviewed did not know what a labor union was, and some said their factory had a "union" that was organized and run by the factory managers. Yet, despite their weak position, these workers strike anyway—wildcat strikes that originate from their own angry demands for justice, much like the uprising of the garment workers in New York a century before. This is an essential point to understand about the realities of the global system. Young people in developing nations are leading thousands of such worker-led strikes, without any guidance from an organized union or other outsiders. They occur because these young people know they are being exploited (though they seldom use that word to describe their conditions). The next time you hear an economist or other experts explain that sweatshops are actually good for the poor people of the world and that any criticism of their working conditions endangers their jobs, ask about the strikes. If the workers are so grateful and happy with their situation, why do they launch so many wildcat strikes in protest? You never read much about this in the American press, but courageous workers in China even mobilize strikes, even though the penalty for the leaders can be ten or twenty years in prison.

The meaning, I believe, is quite positive for the future of the world and its gorgeous variety of people. Because the young strikers confirm that there are universal human aspirations—a thirst for personal dignity and self-advancement and just relationships—that exist across the vast boundaries of geography, race, religion, and culture. The college students and other young Americans who are mobilizing now to protest against the overseas sweatshops that manufacture products for brand-name multinationals like Nike are on the right track, because these students, after all, are the ones who buy the shoes and electronic toys that those young "working girls" are assembling in Indonesia or China or elsewhere. The American students—some of them anyway—are also beginning to recognize that social injustices are not confined to Asia or Central America; some of the same dark practices also flourish within the United States.

The students have discovered a moral question that affects their own lives: If I buy the products that afflicted workers make, do I share in the blame for what happens to those workers? This is the same question posed by the young people who died at the Triangle factory, and Americans found they could not avoid answering it.

To

No. 46

No. 50

No. 61

No. 95

No. 103

No. 115

No. 127

JUDGE M. LINN BRUCE (*counsel*): How high can you succesfully combat a fire now?
EDWARD F. CROKER (*Chief, New York City Fire Department*): Not over eighty-five feet.
BRUCE: That would be how many stories of an ordinary building?
CROKER: About seven.
BRUCE: Is this a serious danger?
CROKER: I think if you want to go into the so-called workshops which are along Fifth Avenue and west of Broadway and east of Sixth Avenue, twelve, fourteen or fifteen story buildings they call workshops, you will find it very interesting to see the number of people in one of these buildings with absolutely not one fire protection, without any means of escape in case of fire.

—Before the New York State Assembly Investigating Committee on Corrupt Practices and Insurance Companies Other Than Life, City Hall, New York City, December 28, 1910

CONTENTS

ILLUSTRATIONS

Following Page 128

All of the photographs in this book are from the UNITE Archives in the Kheel Center for Labor-Management Documentation and Archives at Cornell University, Ithaca, New York, 14853-3901. We are especially grateful to Barbara Morley for assistance with the selection of the photographs and with the captions.

For a website on the Triangle fire, see
http://www.ilr.cornell.edu/trianglefire.

For a website on sweatshop issues, see
http://www.behindthelabel.org.

PART ONE

1. FIRE

I intend to show Hell.
—DANTE: *Inferno,*
CANTO XXIX:96

THE FIRST TOUCH of spring warmed the air.

It was Saturday afternoon—March 25, 1911—and the children from the teeming tenements to the south filled Washington Square Park with the shrill sounds of youngsters at play. The paths among the old trees were dotted with strollers.

Genteel brownstones, their lace-curtained windows like drooping eyelids, lined two sides of the 8-acre park that formed a sanctuary of green in the brick and concrete expanse of New York City. On the north side of the Square rose the red brick and limestone of the patrician Old Row, dating back to 1833. Only on the east side of the Square was the almost solid line of homes broken by the buildings of New York University.

The little park originally had been the city's Potter's Field, the final resting place of its unclaimed dead, but in the nineteenth century Washington Square became the city's most fashionable area. By 1911 the old town houses stood as a rear guard of an aristocratic past facing the invasions of industry from Broadway to the east, low-income groups from the crowded streets to the south, and the first infiltration of artists and writers into Greenwich Village to the west.

Dr. D. C. Winterbottom, a coroner of the City of New

York, lived at 63 Washington Square South. Some time after 4:30, he parted the curtains of a window in his front parlor and surveyed the pleasant scene.

He may have noticed Patrolman James P. Meehan of Traffic B proudly astride his horse on one of the bridle paths which cut through the park.

Or he may have caught a glimpse of William Gunn Shepherd, young reporter for the United Press, walking briskly eastward through the Square.

Clearly visible to him was the New York University building filling half of the eastern side of the Square from Washington Place to Waverly Place. But he could not see, as he looked from his window, that Professor Frank Sommer, former sheriff of Essex County, New Jersey, was lecturing to a class of fifty on the tenth floor of the school building, or that directly beneath him on the ninth floor Professor H. G. Parsons was illustrating interesting points of gardening to a class of forty girls.

A block east of the Square and parallel to it, Greene Street cut a narrow path between tall loft buildings. Its sidewalks bustled with activity as shippers trundled the day's last crates and boxes to the horse-drawn wagons lining the curbs.

At the corner of Greene Street and Washington Place, a wide thoroughfare bisecting the east side of the Square, the Asch building rose ten floors high. The Triangle Shirtwaist Company, largest of its kind, occupied the top three floors. As Dr. Winterbottom contemplated the peaceful park, 500 persons, most of them young girls, were busily turning thousands of yards of flimsy fabric into shirtwaists, a female bodice garment which the noted artist Charles Dana Gibson had made the sartorial symbol of American womanhood.

One block north, at the corner of Greene Street and Waverly Place, Mrs. Lena Goldman swept the sidewalk in front of her small restaurant. It was closing time. She knew the girls who worked in the Asch building well for many of them were her customers.

Dominick Cardiane, pushing a wheelbarrow, had stopped for a moment in front of the doors of the Asch building

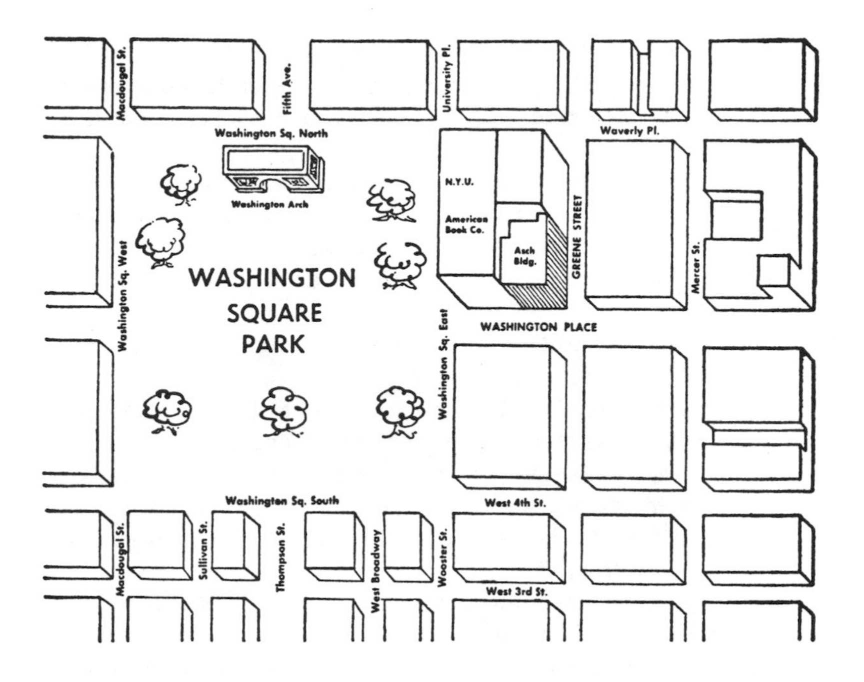

freight elevator in the middle of the Greene Street block. He heard a sound "like a big puff," followed at once by the noise of crashing glass. A horse reared, whinnied wildly, and took off down Greene Street, the wagon behind it bouncing crazily on the cobblestones.

Reporter Shepherd, about to cross from the park into Washington Place, also heard the sound. He saw smoke issuing from an eighth-floor window of the Asch building and began to run.

Patrolman Meehan was talking with his superior, Lieutenant William Egan. A boy ran up to them and pointed to the Asch building. The patrolman put spurs to his horse.

Dr. Winterbottom saw people in the park running toward Washington Place. A few seconds later he dashed down the stoop carrying his black medical bag and cut across the Square toward Washington Place.

Patrolman Meehan caught up with Shepherd and passed him. For an instant there seemed to be no sound on the street except the urgent tattoo of his horse's hoofbeats as

Meehan galloped by. He pulled up in front of 23 Washington Place, in the middle of the block, and jumped from the saddle.

Many had heard the muffled explosion and looked up to see the puff of smoke coming out of an eighth-floor window. James Cooper, passing by, was one of them. He saw something that looked "like a bale of dark dress goods" come out of a window.

"Some one's in there all right. He's trying to save the best cloth," a bystander said to him.

Another bundle came flying out of a window. Halfway down the wind caught it and the bundle opened.

It was not a bundle. It was the body of a girl.

Now the people seemed to draw together as they fell back from where the body had hit. Nearby horses struggled in their harnesses.

"The screams brought me running," Mrs. Goldman recalled. "I could see them falling! I could see them falling!"

John H. Mooney broke out of the crowd forming on the sidewalk opposite the Asch building and ran to Fire Box 289 at the corner of Greene Street. He turned in the first alarm at 4:45 P.M.

Inside the Asch building lobby Patrolman Meehan saw that both passenger elevators were at the upper floors. He took the stairs two steps at a time.

Between the fifth and sixth floors he found his way blocked by the first terrified girls making the winding descent from the Triangle shop. In the narrow staircase he had to flatten himself against the wall to let the girls squeeze by.

Between the seventh and eighth floors he almost fell over a girl who had fainted. Behind her the blocked line had come to a stop, the screaming had increased. He raised her to her feet, held her for a moment against the wall, calming her, and started her once again down the stairs.

At the eighth floor, he remembers that the flames were within 8 feet of the stairwell. "I saw two girls at a window on the Washington Place side shouting for help and waving their hands hysterically. A machinist—his name was Brown—

helped me get the girls away from the window. We sent them down the stairs."

The heat was unbearable. "It backed us to the staircase," Meehan says.

Together with the machinist, he retreated down the spiral staircase. At the sixth floor, the policeman heard frantic pounding on the other side of the door facing the landing. He tried to open the door but found it was locked. He was certain now that the fire was also in progress on this floor.

"I braced myself with my back against the door and my feet on the nearest step of the stairs. I pushed with all my strength. When the door finally burst inward, I saw there was no smoke, no fire. But the place was full of frightened women. They were screaming and clawing. Some were at the windows threatening to jump."

These were Triangle employees who had fled down the rear fire escape. At the sixth floor, one of them had pried the shutters open, smashed the window and climbed back into the building. Others followed. Inside, they found themselves trapped behind a locked door and panicked.

As he stumbled back into the street, Meehan saw that the first fire engines and police patrol wagons were arriving. Dr. Winterbottom, in the meantime, had reached Washington Place. For a moment he remained immobilized by the horror. Then he rushed into a store, found a telephone, and shouted at the operator, "For God's sake, send ambulances!"

The first policemen on the scene were from the nearby Mercer Street Station House of the 8th Precinct. Among them were some who had used their clubs against the Triangle girls a year earlier during the shirtwaist makers' strike.

First to arrive was Captain Dominick Henry, a man inured to suffering by years of police work in a tough, two-fisted era. But he stopped short at his first view of the Asch building. "I saw a scene I hope I never see again. Dozens of girls were hanging from the ledges. Others, their dresses on fire, were leaping from the windows."

From distant streets came piercing screams of fire whistles, the nervous clang of fire bells. Suddenly, they were sounding from all directions.

In the street, men cupped their hands to their mouths, shouting, "Don't jump! Here they come!" Then they waved their arms frantically.

Patrolman Meehan also shouted. He saw a couple standing in the frame of a ninth-floor window. They moved out onto the narrow ledge. "I could see the fire right behind them. I hollered, 'Go over!' "

But nine floors above the street the margin of choice was as narrow as the window ledge. The flames reached out and touched the woman's long tresses. The two plunged together.

In the street, watchers recovering from their first shock had sprung into action. Two young men came charging down Greene Street in a wagon, whipping their horses onto the sidewalk and shouting all the time, "Don't jump!" They leaped from the wagon seat, tore the blankets from their two horses, and shouted for others to help grip them. Other teamsters also stripped blankets, grabbed tarpaulins to improvise nets.

But the bodies hit with an impact that tore the blankets from their hands. Bodies and blankets went smashing through the glass deadlights set into the sidewalk over the cellar vault of the Asch building.

Daniel Charnin, a youngster driving a Wanamaker wagon, jumped down and ran to help the men holding the blankets. "They hollered at me and kicked me. They shouted, 'Get out of here, kid! You want to get killed?' "

One of the first ambulances to arrive was in the charge of Dr. D. E. Keefe of St. Vincent's Hospital. It headed straight for the building. "One woman fell so close to the ambulance that I thought if we drove it up to the curb it would be possible for some persons to strike the top of the ambulance and so break their falls."

The pump engine of Company 18, drawn by three sturdy horses, came dashing into Washington Place at about the same time. It was the first of thirty-five pieces of fire-fighting apparatus summoned to the scene. These included the Fire

Department's first motorized units, ultimately to replace the horses but in 1911 still experimental.

Another major innovation being made by the Fire Department was the creation of high-water-pressure areas. The Asch building was located in one of the first of these. In such an area a system of water-main cutoffs made it possible to build up pressure at selected hydrants. At Triangle, the Gansevoort Street pumping station raised the pressure to 200 pounds. The most modern means of fighting fires were available at the northwest corner of Washington Place and Greene Street.

A rookie fireman named Frank Rubino rode the Company 18 pump engine, and he remembers that "we came tearing down Washington Square East and made the turn into Washington Place. The first thing I saw was a man's body come crashing down through the sidewalk shed of the school building. We kept going. We turned into Greene Street and began to stretch in our hoses. The bodies were hitting all around us."

When the bodies didn't go through the deadlights, they piled up on the sidewalk, some of them burning so that firemen had to turn their hoses on them. According to Company 18's Captain Howard Ruch the hoses were soon buried by the bodies and "we had to lift them off before we could get to work."

Captain Ruch ordered his men to spread the life nets. But no sooner was the first one opened than three bodies hit it at once. The men, their arms looped to the net, held fast.

"The force was so great it took the men off their feet," Captain Ruch said. "Trying to hold the nets, the men turned somersaults and some of them were catapulted right onto the net. The men's hands were bleeding, the nets were torn and some caught fire."

Later, the Captain calculated that the force of each falling body when it struck the net was about 11,000 pounds.

"Life nets?" asked Battalion Chief Edward J. Worth. "What good were life nets? The little ones went through life nets, pavement, and all. I thought they would come down one at a time. I didn't know they would come down with arms entwined—three and even four together." There was one who seemed to have survived the jump. "I lifted her out and said, 'Now go right across the street.' She walked ten feet—and dropped. She died in one minute."

The first hook and ladder—Company 20—came up Mercer Street so fast, says Rubino, "that it almost didn't make the turn into Washington Place."

The firemen were having trouble with their horses. They weren't trained for the blood and the sound of the falling bodies. They kept rearing on their hind legs, their eyes rolling. Some men pulled the hitching pins and the horses broke loose, whinnying. Others grabbed the reins and led them away.

The crowd began to shout: "Raise the ladders!"

Company 20 had the tallest ladder in the Fire Department. It swung into position, and a team of men began to crank its lifting gears. A hush fell over the crowd.

The ladder continued to rise. One girl on the ninth floor ledge slowly waved a handkerchief as the ladder crept toward her.

Then the men stopped cranking. The ladder stopped rising.

The crowd yelled in one voice: "Raise the ladders!"

"But the ladder had been raised," Rubino says. "It was raised to its fullest length. It reached only to the sixth floor."

The crowd continued to shout. On the ledge, the girl stopped waving her handkerchief. A flame caught the edge of her skirt. She leaped for the top of the ladder almost 30 feet below her, missed, hit the sidewalk like a flaming comet.

Chief Worth had arrived at the scene at 4:46½, had ordered the second alarm to be transmitted at 4:48. Two more alarms were called, one at 4:55 and a fourth at 5:10.

In the first two minutes after his arrival, the Chief had assessed the situation. He directed his men to aim high water pressure hoses on the wall above the heads of those trapped on the ledge. "We hoped it would cool off the building close to them and reassure them. It was about the only reassurance we could give. The men did the best they could. But there is no apparatus in the department to cope with this kind of fire."

The crowd watched one girl on the ledge inch away from

the window through which she had climbed as the flames licked after her. As deliberately as though she were standing before her own mirror at home, she removed her wide-brimmed hat and sent it sailing through the air. Then slowly, carefully, she opened her handbag.

Out of it she extracted a few bills and a handful of coins—her pay. These she flung out into space. The bills floated slowly downward. The coins hit the cobblestones, ringing as she jumped.

Three windows away one girl seemed to be trying to restrain another from jumping. Both stood on the window ledge. The first one tried to reach her arm around the other.

But the second girl twisted loose and fell. The first one now stood alone on the ledge and seemed oblivious to everything around her. Like a tightrope walker, she looked straight ahead and balanced herself with her hands on hips hugging the wall.

Then she raised her hands. For a moment she gestured, and to the staring crowd it seemed as if she were addressing some invisible audience suspended there before her. Then she fell forward.

They found her later, buried under a pile of bodies. She was Celia Weintraub and lived on Henry Street. Life was still in her after two hours in which she had lain among the dead.

William Shepherd, the United Press reporter and the only newspaperman on the scene at the height of the tragedy, had found a telephone in a store and dictated his story as he watched it happen through a plate-glass window. He counted sixty-two falling bodies, less than half the final total.

"Thud—dead! Thud—dead! Thud—dead!" Shepherd began his story. "I call them that because the sound and the thought of death came to me each time at the same instant."

As he watched, Shepherd saw "a love affair in the midst of all the horror.

"A young man helped a girl to the window sill on the ninth floor. Then he held her out deliberately, away from the building, and let her drop. He held out a second girl the same way and let her drop.

"He held out a third girl who did not resist. I noticed that.

They were all as unresisting as if he were helping them into a street car instead of into eternity. He saw that a terrible death awaited them in the flames and his was only a terrible chivalry."

Then came the love affair.

"He brought another girl to the window. I saw her put her arms around him and kiss him. Then he held her into space—and dropped her. Quick as a flash, he was on the window sill himself. His coat fluttered upwards—the air filled his trouser legs as he came down. I could see he wore tan shoes.

"Together they went into eternity. Later I saw his face. You could see he was a real man. He had done his best. We found later that in the room in which he stood, many girls were burning to death. He chose the easiest way and was brave enough to help the girl he loved to an easier death."

Bill Shepherd's voice kept cracking. But he was first of all a newspaper reporter, and he steeled himself to see and to report what untrained eyes might miss.

He noticed that those still in the windows watched the others jump. "They watched them every inch of the way down." Then he compared the different manner in which they were jumping on the two fronts of the Asch building.

On the Washington Place side they "tried to fall feet down. I watched one girl falling. She waved her arms, trying to keep her body upright until the very instant she struck the sidewalk."

But on the Greene Street side "they were jammed into the windows. They were burning to death in the windows. One by one the window jams broke. Down came the bodies in a shower, burning, smoking, flaming bodies, with disheveled hair trailing upward. These torches, suffering ones, fell inertly.

"The floods of water from the firemen's hoses that ran into the gutter were actually red with blood," he wrote. "I looked upon the heap of dead bodies and I remembered these girls were the shirtwaist makers. I remembered their great strike of last year in which these same girls had demanded more sanitary conditions and more safety precautions in the shops. These dead bodies were the answer."

FIRE

At 4:57 a body in burning clothes dropped from the ninth floor ledge, caught on a twisted iron hook protruding at the sixth floor. For a minute it hung there, burning. Then it dropped to the sidewalk. No more fell.

2. TRAP

With sad announcement of impending doom.
—CANTO XIII:12

"MY BUILDING is fireproof," Joseph J. Asch insisted.

He talked to reporters in the sitting room of his suite in the Hotel Belmont, a grave-looking man of fifty, sun-tanned and sporting a white mustache. He and his wife had just arrived in the city en route from St. Augustine, Florida, to their home in Saugatuck, Connecticut. He first learned of the fire when he read about it in the Sunday morning papers.

"I was overcome by the horror of it," he emphasized for the reporters. "The architects claimed my building was ahead of any other building of its kind which had previously been constructed."

First plans for the construction of a building at Greene Street and Washington Place were filed in April, 1900, by an architect named John Wooley. He was acting for Ole H. Olsen of the Bronx who owned a 25- by 100-foot lot on the site and had acquired from Asch the adjoining 75- by 100-foot lot.

But Asch changed his mind, took over the combined site from Olsen, and decided to build for himself. His architect, Julius Franke, filed new plans with the Building Department on April 28, 1900. The plans were finally approved on July 13 of that year. The Asch building, representing a cost of about $400,000, was completed January 15, 1901.

The structure was 135 feet high. At 150 feet—or with one additional story—it would have had, as required by law, metal trim, metal window frames, and stone or concrete floors. At 135 feet its wooden trim, wooden window frames, and wooden floors were legal.

The law required only a single staircase in a building in which the floor space at each level was less than 2,500 square feet; if the single floor measured more than that but less than 5,000 square feet the building was required to have two staircases; there would be an additional stairway for each additional 5,000 square feet.

By this measure, the Asch building with an interior area of 10,000 square feet per floor should have had three staircases. This flaw was noted by Rudolph P. Miller, then an inspector for the Building Department over which he had risen to be director by 1911. On May 7, 1900, he wrote to architect Franke that "an additional continuous line of stairway should be provided."

The architect asked for an exception because "the staircases are remote from each other and, as there is a fire escape in the court, it practically makes three staircases, which in my opinion is sufficient."

Miller also insisted that the "fire escape in the rear must lead down to something more substantial than a skylight." The architect replied with the promise that "the fire escape will lead to the yard and an additional balcony will be put in where designated on the plan."

Both staircases were in vertical wells with their steps winding once around a center between floors. The steps were of slate set in metal and measured 2 feet 9 inches in width but were tapered at the turns. The walls were of terra cotta.

Only the Greene Street staircase, with windows between floors facing the backyard, had an exit to the roof. The windowless Washington Place stairs ended at the tenth floor.

At each floor, a wooden door with a wired glass window, opened into the loft. Section 80 of the State Labor Law required that factory doors "shall be so constructed as to open outwardly, where practicable, and shall not be locked, bolted or fastened during working hours."

But in the Asch building, the last step at each landing was only one stair's width away from the door. Therefore it was not "practicable" for the doors to open outward. Therefore all of them opened in.

The Asch building's "third staircase"—its single fire escape —ended at the second floor, despite Miller's requests. But, as Manhattan Borough President George McAneny pointed out after the fire, the law "doesn't compel any sort of building to have fire escapes. It leaves enormous discretionary power with the Building Department."

The dangers implicit in this situation were underscored by Arthur E. McFarlane, an expert in fire prevention and fire insurance. He charged that in many instances speculative builders "decided to build their loft buildings without any fire escapes at all. Others put them in the air shaft which, in case of fire, becomes its natural flue. Others bolted on the antique, all but vertical, 18-inch ladder escapes such as could not legally have been placed upon even a three-story tenement house."

The City's Board of Aldermen was aware of the danger. In 1909 it spent time studying the problem and ended up with proposed revisions of the building code, one of which would have required street-side fire escapes on buildings of the Asch type.

The entire effort, however, was tripped up by a fight that started with rival interests controlling the production and sale of fireproof materials. The Board divided along parallel lines and stalemated all action on the matter.

"I have never received any request or demand from any department or bureau for alteration to the building nor has any request or demand been received by me for additional fire escapes nor has the fire escape on the building ever been unfavorably criticized to me by any official," Mr. Asch continued in his Belmont Hotel suite. And, slicing the air with his right hand, he added, "I never gave the matter any thought."

There was no law requiring fire sprinklers in New York City factory buildings.

Fire Chief Edward F. Croker argued that there had never been a loss of life in a sprinkler-equipped building. These devices, attached to the ceiling, had heat fuses which automatically could set off an alarm and at the same time release heavy sprays of water in the area where heat from a fire had accumulated.

Chief Croker admitted that installation of sprinklers would add about 4 per cent to construction costs. But he stressed his opinion that "sprinklers increase the renting value of a building and so decrease the price of insurance as to pay for themselves within five years."

Sharing this opinion, and on the basis of his own investigations, Fire Commissioner Rhinelander Waldo had ordered sprinklers to be installed in a number of warehouses. Three weeks before the Triangle conflagration, the Protective League of Property Owners held an indignation meeting. The League's counsel, Pendleton Dudley, then issued a statement charging that the Fire Department was seeking to force the use of "cumbersome and costly" apparatus.

The League insisted that the order was arbitrary and imposed a burden of unnecessary expense and that the action was unreasonable, mischievous, and misleading, the *Tribune* reported. And the *Herald* noted that the owners claimed the order amounted to "a confiscation of property and that it operates in the interest of a small coterie of automatic sprinkler manufacturers to the exclusion of all others."

Chief Croker restrained his anger. "If the manufacturers of certain sprinkler systems have formed a combination," he replied, "there are other sprinkler systems that are not controlled by a combination. Nine sprinklers in all have been tested and approved by the National Board of Underwriters."

There was no law requiring fire drills to be held in New York City factory buildings.

In the fall of 1910, the New York Joint Board of Sanitary

Control, a labor-management group that included a number of cooperating public-spirited citizens, investigated work conditions in 1,243 coat and suit shops in the city. The Board had been created after a strike by the International Ladies' Garment Workers' Union in the summer of 1910.

Nine days before the Triangle fire—on March 16—the New York *Call* published excerpts from the report of Dr. George M. Price, the Board's director. Copies of the full report on the investigation had already been sent to the Building, Fire, and Police Departments. In addition, the Board's secretary, Henry Moskowitz, had sent a long list of shops with hazards to Mayor William J. Gaynor.

Dr. Price noted that the coat and suit shops were not the worst offenders in the matter of safety. "Yet," he declared, "our investigation into the conditions in these shops clearly shows that fire prevention facilities are very much below even the most indispensable precautions necessary."

Ninety-nine per cent of the shops were found to be defective in respect to safety: 14 had no fire escapes; 101 had defective drop ladders; 491 had only one exit; 23 had locked doors during the day; 58 had dark hallways; 78 had obstructed approaches to fire escapes; and 1,172, or 94 per cent, had doors opening in instead of out.

Only one had ever had a fire drill.

But at Triangle there had been a warning. In 1909, when the firm was adding to its insurance coverage, P. J. McKeon, an expert and lecturer on fire prevention at Columbia University, was commissioned to make an inspection of the shop.

He was concerned immediately with the crowding of so many people into the top three floors of the building. Upon inquiring, he learned that the firm had never held a fire drill. He noted that without previous instruction on how to handle themselves in such an emergency a fire would panic the girls.

McKeon found that the door to the Washington Place stairway was "usually kept locked," and was told this was because "it was difficult to keep track of so many girls." He thought he had impressed management with the need to hold fire drills. Accordingly, he recommended that Mr. H. F. J. Porter, one of the ablest fire prevention experts in the city, be called in to set up the drills.

On June 19, 1909, Porter wrote to Triangle at McKeon's suggestion, offering to call at management's convenience. He never received a reply to his letter.

There were other portents.

Exactly four months before the Triangle tragedy—on November 25, 1910—fire broke out in an old four-story building at Orange and High Streets in Newark, New Jersey. In minutes, twenty-five factory workers, most of them young women, were dead. Of these, six were burned to death, nineteen jumped to death.

The disaster just across the Hudson River shocked New York and the next day Chief Croker warned:

"This city may have a fire as deadly as the one in Newark at any time. There are buildings in New York where the danger is every bit as great as in the building destroyed in Newark. A fire in the daytime would be accompanied by a terrible loss of life."

Professor Francis W. Aymar of the New York University Law School read Chief Croker's warning. He immediately wrote a letter to the city Building Department saying that from the windows of his classroom he could see the crowded and dangerous conditions in the Asch building across the yard. His letter was acknowledged and assurance of an investigation was given.

Following the Newark fire, the Women's Trade Union League assigned Miss Ida Rauh to study the disaster and draw up a set of conclusions and recommendations. This she did and then, on behalf of the League, Miss Rauh wrote to the January Grand Jury asking to be heard.

Her hearing was short, and fruitless. She was practically dismissed by the foreman of the Grand Jury when she had identified herself. He warned her that "unless you have a complaint of criminal negligence on the part of an official, you had better take your stories to the Corporation Counsel and have him prosecute for violations."

The lesson of the Newark fire was not lost on Alderman Ralph Folks. He introduced a resolution in the Board calling on the Superintendent of the Building Department to investigate and determine if additional legislation were needed to protect the lives of factory workers in New York. Four months before the Triangle fire the resolution was passed.

The day after the Asch building disaster, the *Times* sought out Alderman Folks and asked him what had come of the investigation he had requested in his resolution.

"I don't know. I never heard of it again," he replied.

The *Times* also found Mr. H. F. J. Porter, who had written to Triangle for an appointment on the matter of fire drills.

"There are only two or three factories in the city where fire drills are in use," he declared ruefully. "In some of them where I have installed the system myself, the owners have discontinued it.

"The neglect of factory owners in the matter of safety of their employees is absolutely criminal. One man whom I advised to install a fire drill replied to me: 'Let 'em burn. They're a lot of cattle, anyway.' "

On October 15, 1910—a little more than five months before the tragedy—Fireman Edward F. O'Connor of Engine Company 72 made a routine inspection of the Asch building. Under the definitions of the existing codes, he had to report that the fire escape was "good," the stairways were "good," the building was "fireproof." He noted that an 8- by 10-foot tank on the roof had a capacity of 5,000 gallons of water and that there were 259 water pails distributed over the building's ten floors for emergency use.

In the decade since the construction of the Asch building, about $150,000,000 had gone into the building of similar structures in lower Manhattan. Greater height meant greater returns on land values.

By September, 1909, the greater city had 612,000 employees in 30,000 factories—50,000 more workers than in all of Massachusetts at that time. And, early in 1911, the Women's Trade Union League reported that about half the total number was employed above the seventh floor.

That was just about the height beyond which the finest fire-fighting force in the country could not deal successfully with a fire.

Yet so strong was the feeling of safety in the new buildings that not a single new building or factory law was enacted in the entire decade. Even the number of required stairways, for example, continued to be geared to the area of a single floor with no regard to the number of floors in the structure or the number of workers in the building.

Indeed, the day after the Triangle fire, Albert G. Ludwig, Chief Inspector and Deputy Superintendent of the Building Department, inspected the Asch building. Then he declared: "This building could be worse and come within the requirements of the law."

Joseph J. Asch was right. In his hotel suite, he leaned forward and assured the reporters: "I have obeyed the law to the letter. There was not one detail of the construction of my building that was not submitted to the Building and Fire Departments. Every detail was approved and the Fire Marshal congratulated me."

And that is why, high above the city's streets, beyond the reach of fire-fighting equipment, without benefit of fire sprinklers, proper fire ecapes, or fire drills, 500 employees of the Triangle Shirtwaist Company on that March 25, having heard the bell marking the end of their work day, began to rise from their machines and their work tables with the utmost confidence in their own security and safety.

3. EIGHTH

Now we descend by stairways such as these.
—CANTO XVII:82

"I RANG THE BELL," said Joseph Wexler, eighth-floor watchman for the Triangle Shirtwaist Company. He stood beside the time clock attached to the Greene Street partition, near the single exit on that side of the shop. "I was supposed to look in the girls' pocketbooks, every girl's pocketbook. I got ready. I rang the bell."

What time was it?

Quitting time on Saturday was 4:45 P.M. But Triangle, like Congress, sometimes manipulated its control clock in order to keep working.

On that Saturday, Fire Department Central received the first alarm from Box 289 at 4:45 P.M. At 4:45½ P.M., two telephone calls from 23 Washington Place reported a fire. But the first entry on the matter in the daybook of the Eighth Precinct reads: "Sgt. James Cain and eight patrolmen left for duty at Washington Place fire—4:30 P.M."

What time was it?

It was the time before sundown when bearded Jews in the East Side tenements were beginning to chant the prayer marking the end of the Sabbath. It is called the Havdallah, which means "the great divide." Its joyous words praise the Lord for creating the Sabbath and setting it apart from all the other days of the week.

EIGHTH

Crossing back over "the great divide" and returning to the stretch of ordinary days is marked by the performance of an act forbidden during the holy day. As the sun sets, he who prays pours a small portion of ritual wine into a platter and touches fire to it. The Sabbath ends. And the flames flare up.

Those Jewish workers who had brought a strong religious faith with them to the New World—and had held on to it—paid for their piety. While others remained bent over their sewing machines on Friday afternoon, these stole glances at the sun and when it began to dip, pushed back their chairs, rose from their work and departed—losing time and earnings.

For those on whom the grip of faith had loosened, the sundown meant little. At Triangle, the Sabbath was not set apart. There was work for those who wished to work on Saturdays, and often even on Sundays.

Wexler rang the quitting bell. From where he stood, he could survey the entire shop floor, a huge square measuring about 100 feet on each side. The upper right-hand, northeast, corner of the square was separated from the rest of the shop by a wooden partition, making a vestibule 15 feet long. Behind that partition were the doors to two freight elevators and to the single staircase leading down to Greene Street and up to the roof.

"When we stepped out of the freight elevators in the morning," says Joseph Granick, who worked at Triangle as a garment cutter, "we turned to the right, walked past the staircase door and came into the shop through the single door in the narrow side of the partition. Only one person at a time could pass through that door opening. This is where the watchman stood every night to search the girls' pocketbooks."

Through the door and straight ahead was the back-yard wall. It had eight windows, the two center ones opening onto the single fire escape. Two tables on which cutters worked ran parallel to this northern wall.

The east wall of the shop had ten windows facing Greene Street, one of them behind the partition. Along its length were five more cutting tables, three of them 66 feet long, the other two about half that length, and all of them ending at the Washington Place, southern, wall of the shop. This wall had twelve windows providing plenty of light for the battery of special stitching and sewing machines lined along its length.

The fourth, or west wall, was back to back with the building housing New York University and the American Book Company. Ten feet from where it angled into the Washington Place wall, this side of the square had two passenger elevator doors and to the right of these the single door into the staircase leading to the Washington Place lobby and up to the tenth floor.

A wooden partition to the right of the stairway door enclosed the dressing room, and beyond this were the washrooms. The wall ended with three windows opening on the short wing of the L-shaped back yard across which they faced windows of the University building.

Production of Triangle's shirtwaists began on the cutting tables, where the basic economy of factory methods was achieved. It lay in the ability of the skilled craftsmen to cut many layers of fabric—several dozen at a time—with a single slicing motion.

The tables on which they worked were 40 inches wide, stood 42 inches high, and were separated by aisles that were 30 inches wide. Forty cutters, a closely knit group, were stationed in this department.

Granick remembers that his mother paid one of them, Morris Goldfarb, $25 to get her son a job at Triangle. (Goldfarb was one of Triangle's "toughies" who fought the strikers during the 1909 walkout and helped pass strikebreakers through their lines.)

He had expected to work for nothing at the start. But the $25 had worked magic, and he was hired at $3 a week plus 15 cents an hour for overtime.

Granick was working on lot number 1180 that day, cutting linings and trimmings from a thin fabric called lawn, though sometimes given a fancier name like "longerine." A cutter would pile up many layers of fabric. Then he would place his patterns on top of the pile. This was his main skill. He arranged his patterns to make the shortest possible jig-saw when he first laid them out to get the length for cutting each layer from the roll.

The less yardage he took, the more money he saved for the boss. He had to have a good eye and a strong arm because when he had all of his patterns laid out on top of the layers of fabric, minding the grain of the goods and making certain that no part of the garment was missing, he would start to cut with a short knife that looked like something fishermen use to clean fish. It had a stubby handle and a blade as sharp as a razor. With his left hand the cutter pressed palm down on the pattern while his right hand, grasping the knife, rode around the edge of the pattern, which was bound with metal to prevent nicking.

The tables were boarded up from the floor to about 6 inches below their tops, thus providing large, continuous bins. A long wire with pendant small hooks stretched directly over the full length of each table.

As he sliced out a sleeve or a bodice front, the cutter would set aside the part, hanging the pattern on the overhead wire. From time to time, he would take the cutaway fabric remaining on the table—like dough remaining after the cookies have been stamped out—sweep it together by hand, and fling it through the 6-inch slot on the underside of the table.

There was by-product value for Triangle's proprietors in the cutaways that accumulated under the tables. They were purchased regularly by a dealer named Louis Levy. Between March 25, 1910, and March 25, 1911, he removed the accumulated cutaways six times.

"I waited for an accumulation," said Levy. "The last time I removed the rags before the fire was on January 15, 1911. They came from the eighth floor. Altogether, it was 2,252 pounds."

Granick had received his week's pay and was walking toward the Greene Street exit where Wexler was stationed when he encountered Eva Harris, the sister of Isaac Harris, one of Triangle's two owners. "Eva Harris said she smelled something burning. I looked to the cutting tables. At the second table, through the slot under the top, I saw the red flames."

Eva Harris turned and ran toward Dinah Lifschitz sitting at her desk in the corner of the shop near the three western windows. Dinah handed out the work to the special machine operators on the eighth floor. Next to her desk stood short, stocky Samuel Bernstein, Triangle's production manager, who was related by marriage to both of its owners.

"I heard a cry," Bernstein said. "It was Eva Harris. She was running toward me from the middle of the shop. She was hollering, 'There is a fire, Mr. Bernstein.' I turned around. I saw a blaze and some smoke at the second table from the Greene Street windows. As I ran across the shop toward the fire, some cutters were throwing pails of water."

Another Bernstein, William, a cutter, had grabbed a pail of water that stood near the last window and had thrown it on the fire. "But it didn't do any good," he asserted. "The rags and the table were burning. I went around the partition into the freight elevator vestibule to get another pail of water. But when I tried to go back in, the narrow doorway was blocked with people rushing to the stairs."

Cutter Max Rothen had just hung his patterns on the long overhead wire, having finished the day, when he felt a punch in the back. He turned around and there was Bernstein the manager, "his face white from fright."

"At the same time there were cries of 'fire' from all sides. The line of hanging patterns began to burn. Some of the cutters jumped up and tried to tear the patterns from the line but the fire was ahead of them. The patterns were burning. They began to fall on the layers of thin goods underneath them. Every time another piece dropped, light scraps of burning fabric began to fly around the room. They came down on the other tables and they fell on the machines. Then the line broke and the whole string of burning patterns fell down."

The smoke grew thicker. Those still in the loft began to choke and cough. A half dozen men continued to fling pails of water at the fire.

"But the flames seemed to push up from under the table right to the top," Granick recalls. "I began to look for more water. I thought there might be more pails on the Washington Place side. I began to run there, then I stopped and looked back. I saw in a flash I could never make it. The flames were beginning to reach the ceiling."

Now the flames were everywhere, consuming the fabric and beginning to feed on the wooden floor trim, the sewing tables, the partitions. The heat and the pressure were rising. The windows began to pop, and down in Greene Street, Dominick Cardiane heard a sound "like a big puff."

Samuel Bernstein stood near the flames and cried out for more pails of water. One of the elevator operators came running with a pail. "He left the elevator door open. It made a terrible draft. The wind blew right through the place. We found it was impossible to put the fire out with pails."

Then Bernstein remembered there were hoselines hanging in the stairwells. "I saw Louis Senderman, the assistant shipping clerk from the tenth floor. I hollered, 'Louis, get me a hose!' "

Somehow Senderman managed to reach a hose in the Greene Street stairwell. He fought his way back into the loft, dragging the hose behind him. He reached out the nozzle to Bernstein. "As I took the hose from him, I said, 'Is it open?' But it didn't work. No pressure. No water. I tried it. I opened it. I turned the nozzle one way and then another. It didn't work. I threw it away."

Senderman tried to work the hose. So did Solly Cohen, another shipping clerk. They had as little success as Bernstein. Then a third man tried.

"He was a little fellow, I don't even remember his name, he was an assistant machinist, new in the factory. He came toward me dragging the hose. He handed it to me," Bernstein continued. "I tried it again. I hollered, 'Where is the water? where is the water?' He answered, 'No pressure, nothing coming.'

"Who was the boy? All I know is that he was lost in the

fire. He was pulling me by the hand and screaming. I turned around and I looked at him and the boy was burning. He ran away from me into the smoke."

The machinist, Louis Brown, was in the men's room just north of the dressing room when he heard the cry of "Fire." He dropped the soap with which he was washing his hands and ran into the shop. The first thing he saw was Bernstein standing on one of the tables with a pail of water. He ran to join him.

"When Mr. Bernstein saw me he hollered, 'Brown, you can't do anything here. Try to get the girls out!' I saw the girls all clustered at the door to the Washington Place stairway and I ran in that direction."

Realizing that the fire was now beyond control, Bernstein jumped down from the table and hurried to the Greene Street vestibule, where he concentrated on getting the girls out.

"I wouldn't let them go for their clothes even though it was Spring and many of them had new outfits. One of the girls I slapped across the face because she was fainting. I got her out. I drove them out."

Bernstein worked now at saving the girls with the same drive and force he had used to work those same girls for the benefit of their employers.

On his way to the Washington Place door, Brown had opened the back-yard window leading to the fire escape. Rose Reiner had come screaming out of the dressing room, where she had been giving herself a last inspection in the full-length mirror before going home, when she heard the cries of "Fire!"

"I saw Dinah and she shouted I should go to the fire escape. As I climbed through the window I saw Mr. Brown trying to open the Washington Place door. I went out onto the fire escape."

Filled with terror, Rose slipped and stumbled down the narrow, slatted stairs from floor to floor until she followed the girl ahead of her back into the building through the smashed window at the sixth floor. "I went down. More I don't remember."

She was one of the "clawing" women Patrolman Meehan

found behind the locked door he broke down.

The flames crept nearer to her desk, but Dinah Lifschitz held her post. On her desk was a telephone and a telautograph—a duplicating script writer for sending messages—that had recently been installed by the firm. Both connected directly with the executive offices on the tenth floor.

"I right away sent a message to the tenth floor on the telautograph," Dinah said. "But they apparently didn't get the message because I didn't get an answer.

"Then I used the telephone. I called the tenth floor and I heard Mary Alter's voice on the other end. I told her there was a fire on the eighth floor, to tell Mr. Blanck. 'All right, all right,' she answered me."

Dinah stayed at the phone. She shouted to the switchboard operator to connect with the ninth floor. She was getting no answer.

But her call had given the alarm to the people on the tenth floor. They began to ring for the passenger elevators, and the operators responded to the buzzing from the executive floor. The cars began to pass the eighth floor.

The girls crushed against the eighth floor elevator doors could see the cars going up. "Some of the girls were clawing at the elevator doors and crying, 'Stop! stop! For God's sake, stop,' " Irene Seivos remembered.

"I broke the window of the elevator door with my hands and screamed, 'fire! fire! fire!' It was so hot we could scarcely breathe. When the elevator did stop and the door opened at last, my dress was catching fire."

The car could not hold all who tried to crowd in. Irene Seivos jumped on top of the girls already in the car just as she saw the door closing. Someone grabbed her long hair and tried to pull her out, but she kicked free and rode to safety.

One of those on whom she landed was Celia Saltz. When the fire had started, she was still at her machine. "All I could think was that I must run to the door. I didn't know there was a fire escape. I even forgot that I had a younger sister working with me.

"The door to the staircase wouldn't open. We pushed to

the passenger elevators. Everybody was pushing and screaming. When the car stopped at our floor I was pushed into it by the crowd. I began to scream for my sister. I had lost her, I had lost my sister."

Celia fainted in the car, but in the crush remained on her feet. When she regained consciousness she was stretched out on the floor of a store across the street from the Asch building. "I opened my eyes and I saw my sister bending over me. I began to cry; I couldn't help it. My sister, Minnie, was only fourteen."

While one group of terrified girls struggled to get into the elevators, another small crowd fought to get through the Washington Place stairway door. Some ran from the elevator to the door. Then back again.

Josephine Nicolosi recalls that when she reached the door to the stairs some thirty girls were there. "They were trying to open the door with all their might, but they couldn't open it. We were all hollering. We didn't know what to do. Then Louis Brown hollered, 'Wait, girls, I will open the door for you!' We all tried to get to one side to let him pass."

Crushed against the door was Ida Willinsky, exerting all of her strength in a futile effort to push the crowd back. "All the girls were falling on me and they squeezed me to the door. Three times I said to the girls, 'Please, girls, let me open the door. Please!'

"But they did not listen to me. I tried to keep my head away from the glass in the door. Then Mr. Brown came and began to push the girls to the side."

Brown remembered that when he reached the area of the door he found all the girls screaming. "I tried to get through the crowd. I pushed my way through and tried by main strength to scatter them. But they were so frantic they wouldn't let me through. As I tried to push them to the side, they pushed back. At such a time, a thousand and one thoughts go through your mind. All I could think was 'Why don't they let me through? Why don't they understand that I am trying to get them out?'

"I finally got to the door. There was a key always sticking

in that door and I naturally thought that they must have locked the door. So I turned, all I tried to do was to turn the key in the lock. But the key wouldn't turn to unlock the door. It did not turn. So I pulled the door open. It didn't open right away.

"I had to push the girls away from the door. I couldn't open it otherwise. They were packed there by the door, you couldn't get them any tighter. I pulled with all my strength. The door was open a little while I was pulling. But they were all against the door and while I was pulling to open it they were pushing against it as they tried to get out.

"They were closing the door by their pushing and I had to pull with all my might to get it open."

Brown finally got the door open, and the screaming girls squeezed themselves into the narrow, spiral staircase, pushing and falling in their fright. Brown tried to calm them as he stood at the door. Then the line downward seemed blocked. Squeezing his way halfway down to the seventh floor, he found Eva Harris slumped on the stairs in a faint. At this moment, Patrolman Meehan came panting up the stairs, lifted the girl against the wall, brought her round, and sent her down.

Sylvia Riegler was right behind Brown when he helped pick up Eva Harris. She had been in the dressing room when the fire started. She had just put on her wide velvet hat when her friend, Rose Feibush, ran into the room screaming, took her by the hand, pulled her into the shop, and began to drag her toward a window.

"I saw men pouring water on the fire at the cutting tables. The wicker baskets where the lace runners worked were beginning to burn. I was scared. Rose was pulling me and screaming.

"Suddenly, I felt I was going in the wrong direction. I broke loose. I couldn't go with her to the windows. This is what saved my life. Always, even as a child, even now, I have had a great fear of height.

"I turned back into the shop. Rose Feibush, my beautiful, dear friend, jumped from a window.

"I saw Brown get the door open. Somebody pushed me

through. I don't remember how I got down. I was cold and wet and hysterical. I was screaming all the time.

"When we came to the bottom the firemen wouldn't let us out. The bodies were falling all around. They were afraid we would be killed by the falling bodies. I stood there screaming."

Two men carried Sylvia across the street into a store and "stretched me out on the floor."

She had swallowed so much smoke that they tried to pour milk into her for its emetic effect. But for one who had known hunger, milk could have only one purpose. "They gave me a lot of milk to drink to give me back my strength," is the way Sylvia Riegler remembered it almost half a century later. "But I couldn't hold it. All the time I could see through the store window the burning bodies falling."

On the eighth floor the flames had cut across the shop. Now they rose like a wall, cutting Bernstein and Dinah Lifschitz off from the Washington Place door. The corridor through the flames to the Greene Street door was perilously narrow.

"It was getting dark with smoke and there sat my cousin Dinah trying to get upstairs on the telephone or on the writing machine," Samuel Bernstein said. "She was getting no answer. She screamed 'fire' through the telephone and she screamed it so loud I stopped her. She would have scared the girl on the other end.

"We weren't getting the message through to the ninth floor. Remember, we had to make contact through the tenth floor switchboard. I said, 'For God's sake, those people don't know! How can we make them know?' "

Dinah Lifschitz cried: "I can't get anyone! I can't get anyone!"

Bernstein realized the moment of decision had come.

"I said, 'Dinah, we are the last ones,' and I ordered her to drop the phone and get out. I remembered I had relatives on the ninth floor and they were all very dear to me. I ran through the blaze and the smoke to try to get to the ninth floor.

"I don't know how I got into the Greene Street staircase.

But I could not get into the ninth floor. Twenty feet from the door on that floor was a barrel container of motor oil. I suppose that was burning. I don't know. The blaze was so strong I could not get into the ninth floor. Then I ran up to the tenth floor. I found it was burning there, too."

On the other side of the flames, on the opposite side of the eighth floor, machinist Brown and Patrolman Meehan came back up to the eighth floor to make certain everyone was out.

"We yanked two girls out of a window and got them to the staircase," Brown said. "I went back to the window to see if anyone else was there. The people in the street saw me. They raised their hands and yelled for me not to jump. When I saw that I decided it was time for me to turn around and get out.

"But I couldn't find my way out any more. It was so black with smoke that I couldn't see. I couldn't even see the door. I knew the doorway was about fifteen feet in front of me. I got down on my hands and knees and crawled out that doorway."

Eight floors below, Fireman Oliver Mahoney of Company 72 which had arrived at 4:46½, burst into the Washington Place lobby.

"At first we couldn't even get into the lobby. As I got in, an elevator opened and another group of frightened people got out. We pushed through," said Fireman Mahoney.

Three other firemen carrying the hose on their shoulders followed him up the stairs until he reached the eighth floor landing.

"I had no water with me yet. My men were coming up with the hose. I located the proper place where I was to work. I could see the eighth floor was a mass of flames," Mahoney added.

On the Greene Street side Captain Ruch had dashed up the stairs while his men unrolled the hoses from where they were connected to a hydrant at Waverly Place. He encountered Max Hochfield and restrained him from going back upstairs. Then he returned to the street and led his men, now carrying the hose, up to the sixth-floor level. Here they disconnected the house hose from the standpipe and connected the fire hose.

The Triangle Fire

Captain Ruch carried the hose up to the eighth floor:

"I shouted, 'Start your water.' It came. The fire was so intense it was impossible to stand up. We lay down on our stomachs or on our knees to try to make an entrance. The eighth floor was a mass of fire. . . ."

4. TENTH

A way the margins make that are not burning.
—CANTO XIV:141

THE NERVE CENTER of the Triangle Shirtwaist Company was the suite of executive offices on the tenth floor, which occupied the space along the Washington Place windows. About forty male and female garment pressers worked at tables lined along the Greene Street windows. The entire rear of the floor, facing the back courtyard, was occupied by a large packing and shipping room filled with hanging garments and cardboard and wooden boxes.

Also on the tenth floor was the Triangle switchboard. On that Saturday Mary Alter, a cousin of Mrs. Isaac Harris, added operation of the switchboard to her usual typing chores because the regular board operator was ill.

When Mary Alter first heard Dinah Lifschitz' buzz on the telautograph, she hurried to the machine to receive the mechanical message.

She waited, "but the pen did not move. It simply stuck in the well. As I waited for the pen to start writing, I realized that it was a new machine and that a good many of the girls did not yet know how to operate it properly. They often made mistakes and didn't connect things right. So I went back to my desk thinking that someone was fooling me."

Mary continued typing until the telephone switchboard buzzed.

"It was the eighth floor calling. At first I couldn't make out any clear sound. It was like yelling and I asked, 'What is the trouble down there? what are you yelling about?'

"Then I heard distinctly, 'There is a fire!' So I immediately got up and told Mr. Levine, our bookkeeper, to telephone the Fire Department, which he did. Then I went to tell Mr. Harris and Mr. Blanck that there was a fire on the eighth floor and to see about my father."

Mary Alter's father was the tenth-floor watchman. One of his jobs was to stand guard at the Greene Street exit at going-home time, just as Joseph Wexler was doing on the eighth floor.

Now no one monitored the switchboard. At her eighth-floor desk, Dinah Lifschitz held on to the phone, shouting into its mouthpiece for the ninth floor, getting no answer.

"I did not ring up the ninth floor," said Mary Alter.

The bookkeeper, heeding Mary's order, ran into Mr. Blanck's office, where there was a phone with a direct outside line. "I called Fire Headquarters but they had already received word of the fire so I hung up," he said.

Levine put down the receiver and turned to leave when he saw two children standing in a corner of the office, intently watching him. They were Blanck's two daughters, Henrietta, aged twelve, and Mildred, aged five. Their nursemaid, Mlle. Ehresmann, had brought them downtown for a shopping trip their father had promised. Mrs. Blanck was in Florida with the other children.

Levine told the children to stay in the office where the nursemaid had deposited them while she went to find Blanck. The bookkeeper had seen smoke rising outside the windows and had decided to get back to his own office to put the record books into the safe.

The person in charge of the shipping room was Edward N. Markowitz, and he had also seen the smoke. He was suddenly struck by the realization that on the floor below there were more than 250 persons crowded among the machines and tables. He ran to the Greene Street exit and down the stairs.

Markowitz managed to get through the ninth floor entrance

to the shop just as the first wave of fright had washed over the girls. "They were standing with a sort of dazed look on their faces. They were beginning to push toward the exit and I shouted to them, 'Go nice! There is a fire! Go easy!' I don't know how long I stood in that doorway cautioning the girls. I would put my hand on their shoulders to calm them and I would say, 'One at a time and go to the stairs and get out!' "

The girls in the rear of the crowd began to press forward. Markowitz waved them toward the fire escape near where they stood and shouted to the watchman to guide them through the windows. Then he remembered he had left his order book on the tenth floor.

"That book was very valuable to the firm. I went back to the tenth floor and found it. I had the book in my hand when I turned and saw Mr. Blanck standing there in the middle of the floor. He was holding his little daughter in his arm and he held the older one by the hand. He didn't seem to know which way to turn."

Blanck had been in the shipping room—Markowitz's department—when he heard a cry: "The taxi is here, Mr. Blanck." It reminded him of his promise to the children, and he found them waiting in his office where Levine had made the call to the Fire Department.

Blanck remembered that somebody ran in and said, "Mr. Blanck, there is a little fire on the eighth floor." He left the children and started for the Greene Street side. Then it occurred to him that the children might be frightened, so he went back to get them. "At the front passenger elevator the car came up. All the pressers, all the girls were screaming, 'Save us! save us!' I told the elevator man, 'Take these girls down and come right up again.'

"But while the people were pushing in, my little girl, the five-year old, was swept into the car. I grabbed hold of her hand and pulled. I just got her out of the elevator and I held her close to me.

"The elevator operator took them down, as many as he could, and I stood there for about half a minute, seeing if he would come up again. The minute was too big for me. I started for the other side of the shop. When I saw that I was passing the door to the Washington Place stairs, I turned the handle, thinking I will go down this way.

"I heard Mr. Harris someplace in the shop hollering, 'To the roof! to the roof!' I thought I will be smarter and go down this way. I opened the door. There was so much smoke. I knew the children would not be able to stand the smoke.

"I grabbed the two children and ran as far as the middle of the room, holding them. When I got to the middle of the room, I stopped. The smoke and the flames seemed to be coming from all sides."

As the bewildered Blanck stood clutching his children in the center of the turmoil, up the stairs and through the smoke came Samuel Bernstein, clear-headed and determined to save as many as possible. He had tried to enter the ninth floor only minutes after Markowitz had left it, but by that time it was impossible to get beyond the Greene Street vestibule. The flames had come to the door in the partition.

Breathlessly, Bernstein had then hurried up to the tenth floor. At first glance he saw that "they were all running around like wildcats. I shouted that the only way out was over the roof.

"I saw Louis Silk, a salesman for a textile firm, standing on a table trying to knock out a skylight. Near the table was Mr. Blanck with his two children. One of them was screaming. I went over to Mr. Blanck and told him, 'The only way you can get out is over the roof. But you better be quick about it!' Eddie Markowitz took the younger child and Mr. Blanck held the older one and we began to fight our way out of there."

When Markowitz saw Blanck paralyzed by uncertainty, he dropped his order book and picked up the smaller child. "I pulled him by the coat and I said, 'Come along, Mr. Blanck.' We went right through the loft to the Greene Street stairs. I could feel the flames in back of me. I could feel the heat of them as we went to the roof."

Blanck's partner, Isaac Harris, had posted himself at the Washington Place elevators where he remained to guide the girls until he realized that the elevators seemed to have stopped running.

At this point, "I started to holler, 'Girls, let us go to the roof!' We all rushed to the Greene Street stairs. The smoke was getting heavy and the room was getting dark."

When the girls shrank back from the smoke, Harris urged them on: "Go, one of you, two of you. If you can't all go, better at least one should get out."

Harris led the way up the stairs. Halfway up between the tenth floor and the roof was the window facing the rear yard and through it blew a blast of flame. The girls turned up their coat collars or shielded their faces with their muffs. They reached the roof, their clothing scorched, some with their hair smoldering.

On his way to the roof, Bernstein found Lucy Wesselofsky in a fainting condition. He helped her up.

Lucy was one who had tried to calm the girls "by shouting that everyone would be saved if they would stop trying to pile into the staircase all together.

"In fact, on the tenth floor, where there were about seventy people working, all were saved except one. She was Clotilda Terdanova. She tore her hair and ran from window to window until finally, before anyone could stop her, she jumped out. She was young and very pretty. She was to leave us next Saturday to be married three weeks later."

One of the first to reach the roof was Joseph Flecher, an assistant cashier. He approached the edge of the roof and cautiously peered over the side.

"I looked down the whole height of the building. My people were sticking out of the windows. I saw my girls, my pretty ones, going down through the air. They hit the sidewalk spread out and still."

The roof now became a refuge, an island surrounded by huge waves of smoke and flame. The survivors came staggering out of the structure that covered the Greene Street stairs. They were coughing, screaming, hysterical, and some stumbled perilously close to the edge of the roof.

The adjoining building on the Greene Street side was 13 feet higher than the top of the Asch building. It could be reached by climbing from the top of the staircase covering to the superstructure over the freight elevator shaft.

On the Washington Place side, the New York University–American Book Company building towered 15 feet above the Asch structure. But it too could be scaled by way of the roof over the passenger elevators.

On the tenth floor of the school building, Professor Sommer and the members of his law class had heard the screams of the fire engines. Elias Kanter, one of his students that day, remembers him as a tall, handsome redhead, "ready to adjourn at the end of class to the nearby Brevoort Hotel bar with a handful of students anxious to pursue further some fine point of the law."

Professor Sommer halted his discourse and hurried to the adjoining faculty room which had a window facing the rear courtyard.

"Some of the boys followed me," the Professor said, "and we saw that the ten-story building across the areaway was on fire. The areaway was filled with smoke. We heard ear-piercing shrieks as the girls in the factory appeared at the windows."

The screams of the engines and the girls were also heard in Professor Parsons' horticulture class on the ninth floor. He dismissed his class of forty girls at once, and while they gathered up their notes and books, he also ran to the rear window. "I was shocked by a sight more terrible than I ever could have imagined. I saw a fire escape literally gorged with girls. A great tongue of flame reached out for them."

For a moment, James McCadden, a service worker in the school building, stood beside the professor. "I saw a girl come to the edge of the roof and stand for a minute, looking down. She jumped. Her hair was in flames. I couldn't look any more."

On the floor above, Professor Sommer and George DeWitt, one of the students who had run with him to the rear window, rushed back to the lecture room and summoned the young men to follow them to the roof. There they found ladders left by painters who had been working on the roof during the week.

"We put one across the space between the coping of our building and the skylight above the elevator shaft on the Washington Place side," said DeWitt. Then they lowered a second ladder from the top of the elevator shaft to the Asch roof.

Now they formed a line up to the school roof with Charles Kramer, Frederick Newman, and Elias Kanter starting the slow procession of the frightened women from the Asch roof.

While the students continued their rescue operation on the Washington Place side, others set up a scaling relay at the Greene Street side. There somebody reached down and took Max Blanck's little girl and pulled her up. Then he handed up the older one, and after that he himself climbed up or was pulled up, he didn't remember which.

Harris had led a group from the tenth floor in a rush up the stairs past the open window— "The fire was blowing right in that window"—and once on the roof, he climbed to the adjacent Greene Street building. He and another man "ran to the door on that roof. We found it locked. So we smashed the skylight and hollered for help. A man came up and brought a ladder."

Harris then ran back to the edge of the higher building. Down on the Asch roof he saw Bernstein and Louis Senderman struggling with a salesman named Teschner weighing about 250 pounds who, according to Bernstein, was "shivering like a fish and crying like a baby," and threatening to jump. Together, the three raised him to safety.

At another time Bernstein had "pushed one man up and when he got to the next roof he began to run away instead of staying there and helping the rest of us. I yelled, 'For God's sake, stay and help us push these other people up.' He came back and helped until there were only a few left on the Asch roof.

"I saw the flames were coming right onto the roof on the Greene Street and areaway sides. And nobody was there anymore to push me up. So I ran across the roof all the way to the

Washington Place side where the University is and they pushed me up onto a ladder. When I got near the end of the ladder, I looked down. I saw five or six girls falling from the windows."

In the short wing of the L-shaped back yard, the flames leaped across at the University building. They cracked the glass in the windows and set the facing rooms afire. When the Law Library began to burn, University Vice Chancellor Charles McCracken and other faculty members organized a group of students who rescued the books armful by armful.

From the roof above them, when it seemed that the last Triangle survivor had been rescued, Charles Kramer climbed down the ladder for a final inspection of the Asch roof. He groped through the thick smoke. Flames were now rising on all sides.

He came across the roof carefully. He heard someone moaning and moved in the direction from which the sound seemed to be coming. He found a girl lying across the top steps of the Greene Street stairway, her head on the floor of the roof, her hair smoldering.

"He smothered the sparks in her hair with his hands," the *Sun* reported. He lifted her in his arms and headed back across the roof toward the Washington Place ladders. "Then he tried to carry her up the ladder to the higher roof.

"But because she was unconscious, he had to wrap long strands of her hair around his hands. Dragging her, he slowly made his way up the ladder."

5. NINTH

They closed the portals, those our adversaries.
—CANTO VIII:115

ON THE NINTH FLOOR, the telephone that could have alerted two hundred and sixty persons to the peril boiling up beneath them stood silently on a table in the far, inside corner of the shop. At this table, Mary Leventhal distributed the bundles of cut work brought up from the cutting department on the floor below.

The sewing machine operators who stitched the parts together worked at 240 machines that filled almost the entire area of the ninth floor. These were arranged in sixteen parallel rows running at right angles from the Washington Place wall. Each row had fifteen machines and was 75 feet long. At their far end, the lines of machines left an aisle 15 feet wide in the space between the exit to the Greene Street freight elevators and the windows facing New York University across the back yard.

The machine heads were set into a single work table running the full 75 feet. Every two tables faced each other across a common work trough 10 inches wide and 4 inches deep which connected them. The two rows thus connected formed a plant of 30 machines, and there were eight such plants on the floor.

Under each central trough, at about 8 inches from the floor, a rotating axle, spanning the full 75 feet, drew rotary power from a motor at the Washington Place end of the plant. At that side there was no through passage from aisle to aisle. Only at the far end of the machine lines was such passage possible; all aisles between the plants led only to the passage in the rear.

Fifteen workers sat on each side of the plant, rocking in the rhythm of the work. During working hours, the horizon for each was marked by the workers facing her across the trough, one a little to the right, the other slightly to the left, and by the wicker basket on the floor to her right where she kept the bundle of work to be completed. She sat on a wooden chair; her table was of wood; her machine was well-oiled, its drippings caught and held in a wooden shell just above her knees, and the material she sewed was more combustible than paper.

Each machine head was connected by leather belts to a fly-wheel on the rotating axle from which it drew motive power. The machine operators bent to their work as they fed the fabric parts to the vibrating needle in front of them. Leaning forward, each seemed to be pushing the work to the machine. In turn, the rapid stitching made a rasping whirr as the operator pressed her treadle to draw power, and as if satisfied, the machine passed the work through, sending it sliding and stitched into the center trough.

At her table, pretty, blond Mary Leventhal prepared for the end of the workday by checking the book in which she kept the record of work distributed and work returned by the operators. At her elbow stood the telephone, its yawning mouthpiece mounted on a tubelike standard with the receiver hanging on a hook at the side.

Because Saturday was payday, Mary and Anna Gullo, the forelady, had just finished distributing the pay envelopes. "Mary went back to her table and I went toward the freight elevators where the button was for ringing the quitting bell. I rang the bell," Anna Gullo said.

The machinists pulled the switches and suddenly the rasp of the machines stopped. Now the huge room filled with talk as the operators pushed back from their machines. The chairs

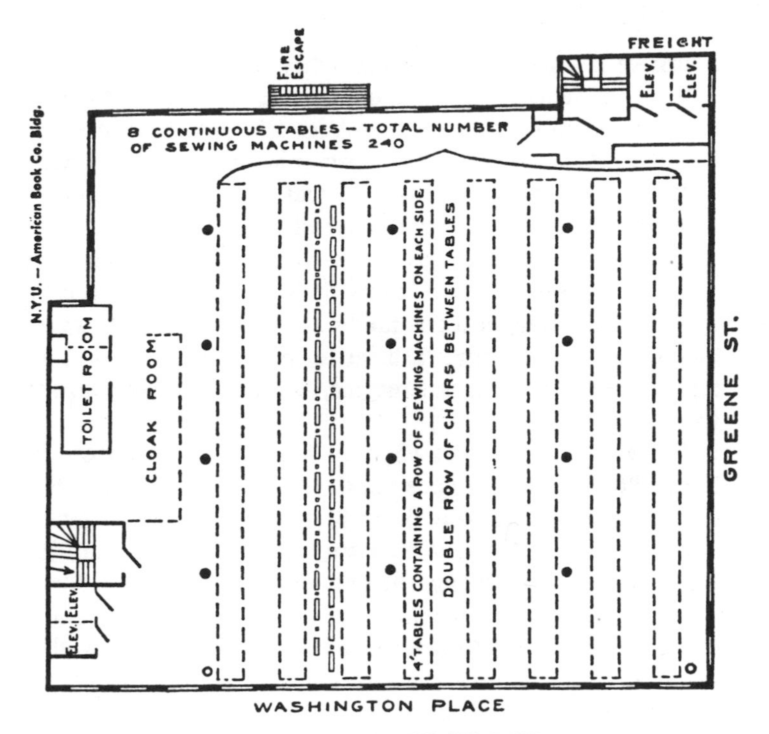

PLAN OF NINTH FLOOR

scraped, locking back to back in the aisles and with the wicker baskets filling the long passageways to the rear of the shop.

The work week was over. Ahead was an evening of shopping or fun and a day of rest. Most of the workers good-naturedly accepted the slow progress up the aisles and around to the dressing room and the exit. Only a few impatient ones jostled ahead to be first out. Max Hochfield had learned that the way to beat the crowd was to avoid the crush in the dressing room and in the rear area of the shop. He kept his coat hanging on a nail protruding on the shop side of the Greene Street partition. When Anna Gullo reached out for the bell button, he was right beside her and out of the door as she rang the signal.

He was the first worker from the ninth floor to learn that

a furnace had flamed up underneath.

When he reached the eighth floor, he could see it was all in flames. Nobody was on the stairs. He ran down another half flight and looked into the courtyard. "I saw people coming down the fire escape. I stopped. I was confused. I had never been in a fire before. I didn't know what to do."

Hochfield continued down the steps, then stopped: his sister Esther was still on the ninth floor. He turned to run back up. "But somebody grabbed me by the shoulder. It was a fireman. I shouted at him, 'I have to save my sister!' But he turned me around and ordered me, 'Go down, if you want to stay alive!' "

This had been the first week for Max and his sister on the ninth floor. On the previous Sunday the Hochfield family had celebrated the engagement of twenty-year-old Esther by giving a party. The guests had visited until early morning, and Max and his sister stayed home the next day.

When they came to work on Tuesday, they found that their machines on the eighth floor had been assigned to two other workers. "Mr. Bernstein told us if we wanted to work we would have to go to the ninth floor. That's how we came to be up there. Maybe if my sister wasn't engaged she would be alive."

The flames invaded the ninth floor with a swiftness that panicked most of the girls but paralyzed others. Pert, pretty Rose Glantz had been one of the first into the dressing room between the door to the Washington Place stairs and the windows facing the University. In high spirits she began singing a popular song, "Every Little Movement Has a Meaning All Its Own."

Some of her friends joined in and when the group finally emerged from the dressing room, giggling and happy, the flames were breaking the first windows on the ninth floor. Laughter turned to screams.

"We didn't have a chance," Rose recalls. "The people on the eighth floor must have seen the fire start and grow. The people on the tenth floor got the warning over the telephone. But with us on the ninth, all of a sudden the fire was all around. The flames were coming in through many of the windows."

Rose ran to the Washington Place stairway door, tried to open it, and when it stayed locked she stood there, screaming. But as the crowd began to thicken, she pushed forward toward the elevator door. "I saw there was no chance at the elevators. I took my scarf and wrapped it around my head and ran to the freight elevator side. I saw the door to the Greene Street stairs was open so I ran through it and down. The fire was in the hall on the eighth floor. I pulled my scarf tighter around my head and ran right through it. It caught fire. I have a scar on my neck."

She made it down the nine floors, meeting the first group of firemen as she neared the freight entrance at street level. There, firemen stopped her from going into the street as they were also doing in the Washington Place lobby with those who had come down from the eighth floor.

Finally, the firemen "escorted us out. I stood in the doorway of a store across the street and watched. I saw one woman jump and get caught on a hook on the sixth floor. I watched a fireman try to save her. I wasn't hysterical any more; I was just numb."

Where was Mary Leventhal? Her telephone did not ring to give the alarm. But Anna Gullo, as soon as she heard the screams of fire, ran across the rear of the shop toward Mary's desk. Then she was swept along with the frantic crowd heading for the Washington Place door.

At the door she tried to exert whatever authority she could command as a forelady. She shouldered the frightened girls aside and tried to open the door.

"The door was locked," she says.

Trapped, she soon shared the terror of those around her. She fought her way out of the crowd and ran to a window on the Washington Place side and tried to open it. The window stuck so she smashed the glass with her hand. Somewhere in the confusion she had picked up a pail of water, and thrown it at the flames.

"But the flames came up higher," Anna says. "I looked

back into the shop and saw the flames were bubbling on the machines. I turned back to the window and made the sign of the cross. I went to jump out of the window. But I had no courage to do it."

Now Anna sought her sister. She headed back across the rear of the shop toward the Greene Street doors, shouting for her sister, Mary.

There were others who cried out for dear ones. Joseph Brenman worked with his two sisters on the ninth floor. Unable to find them, he pushed his way dazedly through the struggling crowd in the rear, calling for the younger of the two, "Surka, where are you, Surka?"

Anna fought her way to the Greene Street door. "I had on my fur coat and my hat with two feathers. I pulled my woolen skirt over my head. Somebody had hit me with water from a pail. I was soaked.

"At the vestibule door there was a big barrel of oil. I went through the staircase door. As I was going down I heard a loud noise. Maybe the barrel of oil exploded. I remember when I passed the eighth floor all I could see was a mass of flames. The wind was blowing up the staircase.

"When I got to the bottom I was cold and wet. I was crying for my sister. I remember a man came over to me. I was sitting on the curb. He lifted my head and looked into my face. It must have been all black from the smoke of the fire. He wiped my face with a handkerchief. He said, 'I thought you were my sister.' He gave me his coat.

"I don't know who he was. I never again found my sister alive. I hope he found his."

Was the stranger Max Hochfield, destined never to see his sister Esther alive again, or was it Joseph Brenman still looking for Surka?

On the ninth floor, those able to make their way up the long, obstructed aisles, pushed into the crowded rear area of the shop. Here, two tides struggled to move in opposite directions. Few knew that behind the nearby shuttered windows was the fire escape that could lead to safety.

Nellie Ventura was one of those who knew. She reached a

small group at the fourth window from the left. The window had been raised, but the outside metal shutter remained firmly closed. Some banged on it with their hands. But two, working on the rusted metal pin holding the shutter closed, succeeded in lifting it. The shutters swung open. Nellie Ventura stepped over the 23-inch-high sill and down to the slatted balcony floor.

She saw thick smoke with tongues of fire at the eighth floor. "At first I was too frightened to try to run through the fire. Then I heard the screams of the girls inside. I knew I had to go down the ladder or die where I was.

"I pulled my boa tight around my face and went. I do not know how I got down to the courtyard at the bottom. Maybe I jumped, maybe somebody carried me. I remember a fireman led me through a hallway and out into the street. At first I couldn't remember where I lived. A policeman took me home."

Panic-stricken girls battled each other on that rickety, terrifying descent. Of her own flight, Mary Bucelli could recall only that "I was throwing them out of the way. No matter whether they were in front of me or coming from in back of me, I was pushing them down. I was only looking out for my own life."

The last to get to safety by way of the fire escape was quick-witted Abe Gordon who, at sixteen, had started working at Triangle as a button puncher. He loved the marvelous sewing machines, the Singers and the Willcox and Gibbs stitchers, and his dream was to become a machinist.

The steppingstone to that job was the assignment as belt boy. Then his task would be to listen for the call of an operator whose machine had lost power because its strap connecting with the fly wheel on the axle had snapped. He would then sidle swiftly down the crowded aisle, creep under the machine table, and make the repair.

Abe saved twelve dollars out of his scant pay and bought a watch fob for the head machinist. In no time at all he was promoted from button puncher to belt boy.

Now, pushing out onto the fire escape, he sensed its inadequacy. "I could hear all the screaming and hollering in back of me. At the sixth floor, I found an open window.

"I stepped back into the building. I still had one foot on the fire escape when I heard a loud noise and looked back up. The people were falling all around me, screaming all around me. The fire escape was collapsing."

Even in the midst of the horror, some failed to grasp its finality. When Yetta Lubitz emerged from the dressing room she was surprised at the sight of the commotion. As a worker "on time" rather than "by the piece," punching out on the time clock had become a most meaningful ritual of her daily routine. She managed somehow to get through the crowd to the time clock, inserted her card in the receiving slot and pushed the punch lever. The ring of its bell seemed to awaken her to the seriousness of the situation around her. She became frightened and ran toward the dressing room.

"Then, all of a sudden, a young dark fellow—he was an operator but I didn't know his name—was running near me. I ran with him to the Washington Place door. He tried to open the door. He said, 'Oh, it is locked, the door is locked.' "

Now panic seized her. She ran back to the dressing room, then out of it to the three windows facing the school building—"the flames had knocked them out and were licking into the shop"—then back again into the dressing room.

"I just stood there crying," Yetta recounted. "The young man, the same dark one, was near me and he snapped, 'Oh, keep quiet; what's the use of crying?' So I felt ashamed and stopped it. But when he was gone, I started to scream again.

"Then I saw the girls were running to the Greene Street side and I started to run, too. The fire was burning in the aisle at the fire escape."

Unable to pass through the rear area, Yetta climbed up onto a machine table. These were 30 inches high, and the machine heads added another 12 inches. Yetta climbed over two plants of machines, past the fire in the rear area. "I looked back and I saw one old Italian woman who couldn't jump down from the machine table. She took a few steps back and forth and she was looking down. But she couldn't get down.

"I covered my face with my skirt and ran into the Greene Street passage. I got to the roof and then it occurred to me that I had forgotten to take my time card out of the clock."

The dressing room, adjacent to the door to the Washington Place stairs, became for some the final place of decision, for others a temporary refuge before death.

Lena Yaller remembered it as being "filled with smoke; everybody was talking or screaming; some in Jewish and some in Italian were crying about their children."

When Ethel Monick looked up from her machine, she saw "fire coming in all around us. I saw women at other machines become frozen with fear. They never moved."

But Ethel, only sixteen, moved. She ran to the Washington Place stairway door, and when she saw it was locked, backtracked into the dressing room. "I was looking for something with which to smash the door. It was wired glass on top.

"In the dressing room men and women were laughing but in a strange way I could not understand at that time. I yelled at them to stop laughing and to help me find an unused machine head to smash the door."

When she left the dressing room, Ethel ran to a window, resolved to jump. "Then I saw in my mind how I would look lying there on the sidewalk and I got ashamed. I moved back from the window." Ethel headed back to the elevators.

The escape routes were closing off. Ida Nelson, clutching her week's pay in her hand, ran to the fire escape windows. She looked out and all she could see was heavy smoke laced with flames.

"I don't know what made me do it but I bent over and pushed my pay into the top of my stocking. Then I ran to the Greene Street side and tried to get into the staircase."

In the few minutes since Anna Gullo had gone down the stairs, that route had been cut off by fire. Now Ida Nelson saw that "I couldn't get through. The heat was too intense.

"I ran back into the shop and found part of a roll of piece

goods. I think it was lawn; it was on the bookkeeper's desk. I wrapped it around and around me until only my face showed.

"Then I ran right into the fire on the stairway and up toward the roof. I couldn't breathe. The lawn caught fire. As I ran, I tried to keep peeling off the burning lawn, twisting and turning as I ran. By the time I passed the tenth floor and got to the roof, I had left most of the lawn in ashes behind me. But I still had one end of it under my arm. That was the arm that got burned."

Rose Cohen had already reached the roof by the time Ida came through the door. Both were sleeve setters, but in the huge ninth-floor factory they had never spoken to each other. However, Rose recalls that "there was one little girl who, like me, saved herself by running to the roof. She had wrapped white goods all around herself and one piece was still burning. I ran to her and helped her beat out the flames. Then I tried to hold back another girl who tore herself away from us and ran back into the stairway to look for her sister."

Rose had been on her way to the dressing room when she heard the cry of fire. She remembers that many of the girls still in the aisles were "caged in by the wicker work baskets." As she ran she tried to down the rising panic in her heart with the thought of what, she says, "all greenhorn immigrants like my parents used to say: 'In America, they don't let you burn.'

"I ran into the dressing room with the machinist and some of the others. The walls of the dressing room began to smoke. We ran back into the shop. Girls were lying on the floor—fainted. People were stepping on them. Other girls were trying to climb over the machines. Some were running with their hair burning.

"I followed the machinist to a window and he smashed the glass with his hand to let the smoke out—it was choking us. Instead, the flames rushed in. For a few seconds I stood at that window. My hair was smoldering. My clothes were torn. I turned and ran to the Greene Street exit.

"I put my hands on my smoldering, long hair and I started up the stairs. On the tenth floor, I went in through the door.

The place was empty. All I could hear was the fire burning. Here, I thought, I would die—here was the end. I didn't realize that right above me—one more floor—was the roof.

"Then I heard some one calling, 'Come to the roof! Come to the roof!' I turned and saw him in a corner near the staircase holding an armful of record books. If not for him I would have died there on the tenth floor. My life was saved by a bookkeeper."

The red flames had raced across the big room, feeding on the flimsy fabric, the wicker baskets, the oil-soaked machines and floor. The machine tables were burning. Solid fire now pushed forward from the rear of the shop, including the Greene Street stairway. It divided the trapped ones into two groups.

One group remained cornered in the areaway in front of the Washington Place elevator doors and staircase. The stairway door remained closed, the girls clawing at it and screaming.

The second group was trapped, scattered, in the aisles between the sewing plants. Now the advancing flames forced them back down the aisles toward the windows facing Washington Place. Some climbed over the rows of machines only to get caught in the last aisle running along the Greene Street windows. The fire backed them into the windows. And in the street the crowds watched helplessly as they plunged into space.

Only the little passenger elevators continued to scoop some of the dying back to life. The Washington Place elevators measured 4 feet 9 inches by 5 feet 9 inches. They were designed to carry about fifteen passengers. Yet Gaspar Mortillalo, one of the two operators, was certain that in his last trips down from the inferno he carried twice that number.

He and Joseph Zito, both in their twenties, had been sitting in their cars at street level, waiting for Triangle—the only firm in the building still working at that hour—to check out. Suddenly, the bells in both cars began to ring insistently. They heard frantic banging up above, the sound of glass smashing, then screams.

"Gaspar didn't say anything to me and I didn't say anything to Gaspar. We both started our cars right up," said Zito.

In these elevators the cable ran through the cage. To start the car moving up, the operator reached up, took a firm grip on the cable and sharply pulled down; he reversed the procedure to descend.

It was Zito's recollection that he must have made seven or eight trips altogether. "Two of them went to the tenth floor, one to the eighth floor and the rest to the ninth. Gaspar and I must have brought down a couple of hundred girls. Twice, I went through smoke and flames that came right into the car.

"When I first opened the elevator door on the ninth floor all I could see was a crowd of girls and men with great flames and smoke right behind them. When I came to the floor the third time, the girls were standing on the window sills with the fire all around them."

Now the elevators became the last link with life. The struggle to get into them became desperate. On their last trips Zito and his partner fought to close the doors before descending.

Fannie Selmanowitz pushed herself into one of the cars as it started to move down. "There wasn't enough room for a pin in that elevator. All the way down I was being pressed through the open door of the car against the side of the shaft."

The last person to get into the last car to leave the ninth floor was Katie Weiner. "I was searching for my sister Rose but I could not find her. Then I found myself crushed against the elevator door, knocking and crying like the others for him to come up.

"He didn't come up. I was choking with the smoke. I went to a window and put my face out to get some fresh air and I calmed down. Suddenly, the elevator came up and the girls rushed to it. I was pushed back to the staircase door. I was crying, 'Girls, girls, help me!' But they kept pushing me back.

"The elevator started to go down. The flames were coming toward me and I was being left behind. I felt the elevator was leaving the ninth floor for the last time. I got hold of the cable that went through the car and swung myself in, landing on the girls' heads.

"All the way down," Katie Weiner remembers, "I was on the people's heads. I was facing downward and my feet were extending out into the shaft. As we went down, my feet were hurting horribly, my ankles were hitting the doors and I was crying, 'Girls, my feet, my feet!' "

Crushed in that elevator on that last trip was Josephine Panno. She screamed all the way down. Through all of the panicked rush to the elevators she had held firmly to the skirt of her daughter, Mrs. Jane Bucalo. But in the final surge that carried her into the car, her grip had broken.

As the elevator began its descent, she caught sight of her daughter's fear-torn face in the crowd being left behind. She screamed and struggled to raise an arm in an effort to reach out to her daughter. But her arms were pinned to her side by the crush.

"When we got to the bottom," she remembered, "girls were crying all around me as they trampled each other to get out. But I could not hear the cries of my daughter." She tried to run up the stairs and fought the policemen in the lobby until she fainted.

Where the elevator had been, there was now only a gaping hole. Sarah Cammerstein stood at the edge and watched the car slip slowly downward. "I thought the elevator was falling. It didn't seem to be driving down but only sinking slowly under too much weight. That's why I hesitated.

"But when I saw it was still at the seventh floor I knew it wasn't falling. I made the decision of my life. I threw my coat down onto the roof of the elevator and jumped for it."

Sarah Cammerstein blacked out. When she opened her eyes again she was lying face up on the roof of the elevator.

"I was alone," she says. "I was spread out on my coat and I looked straight up into the elevator shaft. I could see the flames coming out of the eighth floor. I couldn't move.

"But the elevator was moving. It was going up, straight to the flames. I began to scream. I found the strength to bang my fist on the top of the elevator."

The car stopped. For a moment it was motionless. Then it slowly descended to put its roof on a level with the exit to the lobby and Sarah Cammerstein was lifted out.

Both elevators began their final ascent. The right-hand elevator reached to just below the eighth floor where the heat of the flames pushing into the shaft had bent its tracks. It returned to the lobby. The elevator on the left, Zito's car on which Sarah Cammerstein had been lowered, never rose above the lobby. Falling bodies began to crush in its roof.

Celia Walker stood at the ninth floor opening above Zito's car, afraid of falling, seeking the courage to jump. She was proud of the fact that the other immigrant girls at Triangle considered her "a real Yankee." Celia had come to America with her parents when she was five years old. Bright and zestful, she spoke with what she liked to consider "a true American accent." Her job was to examine finished garments and to return them for cleaning or repair, if needed.

In the terror touched off by the cry of fire, Celia felt panic challenging her usual self-sufficiency. "The girls were climbing over the machine tables. So was I. The aisles were narrow and blocked by the chairs and the baskets which were beginning to burn. I jumped from one table to the next without getting down. I was twenty years old and I could jump.

"The first time I saw the elevator come up, the girls rushed in and it was filled in a second. When it came up again, the girls were all squeezing against the door and the minute it opened, they all rushed in again. This time I thought I was going to be lucky enough to make it, but just as I got to the door, the elevator began to drop down. Somebody in front of me jumped.

"Soon I found myself standing at the edge, trying to hold myself back from falling into the shaft. I gripped the sides of the open door. Behind me the girls were screaming. I could feel them pushing more and more.

"I knew that in a few seconds I would be pushed into the shaft. I had to make a quick decision. I jumped for the center cable. I began to slide down. I remember passing the floor numbers up to five. Then something falling hit me.

"The next thing I knew was when I opened my eyes and looked up into the faces of a priest and a nun. They were trying to help me. I was in a bed in St. Vincent's Hospital.

"They had found me at the bottom of the shaft on the elevator. Others had fallen on top of me. My head was injured. I had a broken arm and a broken finger. Down the middle of my body I felt the burning of the cable which had torn right through my clothes.

"One of the nurses told me she thought it was wonderful that I had enough presence of mind when I jumped to wrap something around my hands in order to save them from injury as I grasped the cable. But it wasn't presence of mind. It was a new fur muff for which I had saved many weeks. Fire or no fire, I wasn't going to lose that muff. I think the right word for my presence of mind is vanity."

Others jumped or fell through the 2½-foot opening into the left elevator shaft. Sarah Friedman had seen the elevator slide away on its last descent. She grabbed the cable at the side of the shaft. "I slid all the way down and ended up on top of the car where I lost consciousness. One of my hands was burned by the friction. When I opened my eyes I was lying in the street—among the dead."

One of the last to jump—and survive—was May Caliandro Levantini, a mother of three children. She tried to stop her fearful, relentless progress toward the elevator pit. Her hands scratched for a hold on the grating alongside the shaft opening.

But when she felt that the pressure behind her could no longer be resisted, she turned and leaped for a cable.

"I was on top of the elevator cage, face up," she said. "I saw firemen going up the stairs with a hose. One of them called to me through the grill as he went up, 'You are all right!' I kept crying, 'Look! Look!' That's all I could say. They couldn't see what I could see way up there in the shaft."

She could see others being pushed out into the shaft at the ninth floor. She rolled over toward the wall to be out of the line of their plunge.

The Triangle Fire

Joseph Zito heard the bodies hit, felt his car shiver with each new impact. "A body struck the top of the elevator and bent the iron. A minute later another one hit.

"The screams from above were getting worse. I looked up and saw the whole shaft getting red with fire. I knew the poor girls up there were trapped. But my car wouldn't work. It was jammed by the bodies."

Then the car slipped down to the bottom of the shaft.

"It was horrible," Zito added. "They kept coming down from the burning floors above. Some of their clothing was burning as they fell. I could see the streaks of fire coming down like flaming rockets."

Now there was no exit. The flames roared louder, steadier. They poured up the Greene Street stairwell. They pushed out of all of the windows. They blew into the Washington Place elevator shaft. In the center of the shop, they billowed as a single deep layer of flame. And the door to the Washington Place staircase held fast and there life screamed to an end.

"I have never been able to forget that maybe I could have saved pretty Mary Levanthal," cutter Joseph Granick says.

"Only a few minutes before the fire she came down to the eighth floor where I was cutting trimmings. She said to me, 'Joe, I have a few girls coming in tomorrow. I need a few dozen cuffs.'

"I gave her five bundles of cuffs. Why didn't I hold her back? Why didn't I talk to her a little longer? Why didn't I argue with her? If she had stayed only a few more minutes she would have escaped with us. But, no! She went back to the ninth floor to die."

6. ESCAPE

For never to return here I believed.
—CANTO VIII:96

THE city came running.

On the East Side, the old women heard the screams of the fire engines, distant and shifting like wild beasts dashing through the labyrinth of narrow streets. They threw their knitted shawls over their heads, wrung their hands, and ran.

The clang and screams of the engines converging from all sides on Washington Place made thousands pause. On nearby Fourteenth Street, Saturday afternoon crowds in front of the department stores and the Academy of Music at Irving Place heard the terrible sounds. In the tenement canyons south of Washington Square, and in the offices along lower Broadway the insistent engine whistles announced an unknown catastrophe.

Wherever they were in the streets, people looked skyward and knew the place of disaster. Faces turned toward the pillar of black smoke steadily rising in the sky. Thousands began to race the screaming engines toward the Asch building.

They came by Third Avenue elevated and by Fifth Avenue double-decker buses. In the more remote reaches of the East Side they boarded streetcars or commandeered rides on horse-drawn wagons.

As they drained toward Washington Place, groups coalesced into crowds overflowing the sidewalks, running in the cobblestoned streets, parting only to let the galloping fire horses dash through. They turned the final corner breathlessly, horrified as they saw the flames, speechless and dumb on first viewing the holocaust, then making the sounds of helpless terror and grief as they saw the bodies on the sidewalk.

But a few ran the other way.

These had escaped the flames in the Asch building. Wild-eyed, disheveled, they had come staggering out of the building, had seen the bodies of friends and co-workers twisted on the sidewalk, and had fled for the sanctuary of home.

Anna Dougherty ran. Her long hair had come loose, her clothing was torn and smoke-smudged. She ran through Broadway near Wanamaker's store, said the *Herald,* "and gave to a large shopping crowd the first intimation of the fire."

Hysterically she darted between the streetcars and the horse-drawn wagons on the busy street, shouting all the time to the people watching her in astonishment:

"Don't let them hurt me! Don't let them hurt me!"

Traffic stopped. Kind hands led her to the sidewalk. She struggled weakly with a policeman who tried to help her. Then she sobbed out her name, told where she lived, and was led away by a group of women.

Dora Appel had come out of the eighth-floor Washington Place staircase door, stumbling downward to the street. There she saw the smashed bodies of her shopmates.

"I couldn't stay there."

She doesn't remember how she got home to Broome Street.

"I walked. I remember coming to my house. I was numb. I walked up the stairs, through the apartment to my bedroom and threw myself across the bed."

There she lay in a sleep of exhaustion. She awoke to the sound of crying on the other side of her bedroom door.

"It was my friend from the next apartment. She was crying for me, pleading to know if I was home. I ran to the door and threw it open. I called to her, 'I am here! I am alive!' Oh, how she ran to me. We were together till early morning. The families around us talked and talked. But we two couldn't stop crying."

Rose Cohen had also sobbed herself to sleep on her bed in the dark bedroom of a long railroad flat on Lewis Street. No one was home when she arrived.

In her sleep she heard shouting and opened her eyes to the darkness. Down the long line of rooms, in the kitchen, her cousin Harry was shouting and crying. He had made the rounds, looking for Rose and had been unable to find her. He feared the worst had happened.

"My mother asked him what had happened. He began to tell her about the fire. I got up from the bed and began the long walk to the kitchen, passing through one room after another as in a dream. Finally, I stood in the kitchen doorway, supporting myself by holding on to the door frame. Then everything broke apart. My mother took one look at me and collapsed to the floor. I began to cry and scream hysterically.

"I couldn't stop crying for hours, for days," says Rose. "Afterwards, I used to dream I was falling from a window, screaming. I remember I would holler to my mother in the dark, waking everybody up, 'Mama! I just jumped out of a window!' Then I would start crying and I couldn't stop."

When Ethel Monick started for home on the East Side, her clothing was torn and soiled, her face was dirty. She reeked with the odor of smoke. In her heart, the terror she had seen contended with the punishment she anticipated.

Her father was "strict." Every morning he doled out fifteen cents to her, "a nickel to go to work, a nickel to come home and a nickel for lunch."

Late and disheveled, she opened the door to the apartment. Before she could speak, her father began to berate her, shouting this was no time, this was no way for a respectable girl to come home.

"I got a licking from my father while my mother stood in a corner respecting his anger. He kept calling me names and I kept crying, 'but, Pa. . . .' He wouldn't listen to me and commanded me to go to bed. I guess while I was sleeping they found out the truth because when I got up they were all standing around me, smiling and kissing me."

Josephine Nicolosi, half blind from shock, had come out of the Washington Place lobby with her friend Frances LoCastra. Both lived as neighbors on Elizabeth Street. Confused, they linked arms and started for home.

"A boy from our block named Frank must have been at the fire. He saw us, took one girl on each arm, led us home. Frank had a reputation of sometimes taking a little too much to drink. And in our neighborhood, if a fellow asked a girl to marry him and she turned him down he would sometimes take terrible revenge.

"We turned into Elizabeth Street, dirty, crying, Frank and us reeling from side to side. The block was quiet. We came straight down the middle of the street.

"My father was in a barber shop in the middle of the block, getting a shave. Somebody ran into the shop and shouted to him: 'Your daughter is coming and her face is all bloody!' My father jumped from the chair. I saw him come tearing out of the barber shop, his face covered with soap lather, his hands twisting a towel.

"The whole block rang with his shouting. 'Who did it? Tell me, who did it? I'll kill him!' he kept roaring."

Josephine knew what was going to happen next. As her father ran to her, the window on the third floor of the house opposite the barber shop shot up.

"Out came my mother's head. She took one look at me from up there, spied the towel in my father's hands and let out a scream that had all the other windows on the block going up in a second.

"She screamed again and disappeared from the window. My father was shaking me by the shoulders hollering, 'I'll kill him! I'll kill him!' Seconds later, my mother caught up with us, crossing herself and crying 'Who? Who?'

"I fell into her arms and while my father continued to threaten and my mother kept asking, all I could do was sob over and over again, 'Tutti morti . . . mama, tutti morti.' "

For some the horror stretched through many nights and

days. Even now, Max Hochfield, who was stopped from turning back up the staircase to look for his sister, remembers how he became obsessed with the idea of avenging her death. He thinks now that the idea came to him during the first night after her death.

"I began to plan how to get a gun," he says. "I would go to collect the wages they owed me—and my sister. The bosses would be there. I would come in and ask for the money. I would kill them."

The trouble with the plan was that until he collected the pay, Hochfield was virtually penniless. "I couldn't buy the gun before I got the pay. And I would have no chance for revenge after I got it."

Then Hochfield had another inspiration. He was a good union man, member of Local 25 of the International Ladies' Garment Workers' Union. He would go to the union office, tell his plan, and borrow money for the gun.

He went.

"I found one of the officers," he recalls. "Nobody else was around. I told him the secret I carried in my heart. I had to avenge my sister's death.

"He listened to me and said, 'No, not that way. I know how you are suffering. But you're a young fellow. You'll ruin your life. Take my advice. Killing won't do you any good and it won't do us any good. Shooting? no; making the union stronger? yes; that's the way.' "

But even now Max Hochfield says: "Still and all, if I had had the money and if I had known where to buy a gun, maybe I would have gone through with the plan."

Like Hochfield, Isidore Wegodner escaped down the Greene Street stairs from the ninth floor, where he and his father had come to work four months earlier as sleeve setters. He was near that exit when he heard the first cry of fire and had no difficulty reaching the street.

But unaware of the extent of the disaster, he had left his father behind. Only when he emerged into the body-littered street did he realize what was happening. The firemen stopped him when he tried to rush back into the building.

They wouldn't even let him cross the street to look among the dead, and he began to cry softly, certain his father lay among them. He moved away, looking into other faces and asking for his father. He spoke only Yiddish, therefore only dimly perceived that there was something called a morgue to which the dead were to be taken.

Suddenly his young heart was lifted by the thought that perhaps his father was seeking him even as he was searching for the old man and that not having found him he had gone home in the expectation of finding him there.

Isidore ran to the Third Avenue elevated, then ran all the way from the 116th Street station to his sixth-floor home on 119th Street. But his father was not there, and when he turned to go down to the street, he lied to his unknowing mother, telling her he had forgotten to buy the old man his newspaper.

In the street, he ran again, determined to return to Washington Place and to find out where this thing called a morgue was located. He missed a train by seconds and stood on the platform breathing hard, watching another pull in on the opposite platform.

"I saw him come out of the train, my dear father who was a quiet man, a dignified man. He looked battered. His pants were torn and in places his flesh showed through. His hat was gone, his face was dirty and bloody. On top of it all he wore a fancy, clean jacket that someone had thrown around his shoulders because his shirt had been ripped off. He stood on the platform dazed and the people walked around him.

"I remember," says Isidore Wegodner, "how with my last strength I shouted to him, how I went tearing over the little bridge that connected the two platforms, how we fell into each other's arms and how the people stopped to look while sobbing he embraced me and kissed me."

7. NIGHT

. . . a little rivulet,
Whose redness makes my hair still stand on end.
—CANTO XIV:77

THE FIRE was brought under control in eighteen minutes. It was practically "all over" in half an hour, said the *Times,* noting that because the building was fireproof, "it showed hardly any signs of the disaster. The walls were as good as ever; so were the floors; nothing was any the worse for the fire except the furniture and the men and girls employed in its upper three stories."

Most of the victims were the main support of their immigrant families, spoke little English, and fled to their homes in the lower East Side as soon as they gained the sidewalk, the paper concluded, adding that it was therefore impossible to estimate the number of those who had escaped the flames.

Shortly after five o'clock, about ten thousand people had been drawn to the neighborhood of the fire. "Great hordes came marching up from the East Side," said the *World,* drawn by the smoke marker in the sky. By seven o'clock the size of the crowd had doubled.

From high water towers and from ten hose lines firemen continued to pour steady streams into the top floors of the building. The water smacked the face of the building with a loud sound; it filled the air with a heavy spray; it came cascading out of the windows, down the sides, out of the lobby, falling like a "miniature Niagara," said the *Times,* and turning the gutter into a blood-stained rivulet.

A pile of thirty bodies lay on the Greene Street sidewalk next to the building. The water soaked them. Surgeons from Bellevue Hospital, dripping wet and cold, burrowed among the bodies, searching each for a spark of life. They dragged tarpaulins along behind them, covering the lifeless heap as they worked across it.

The first medical men to arrive was a group of internes from Bellevue under the direction of Doctors Byrne, Read, and Kempf. Soon they were joined by doctors from New York and St. Vincent's Hospitals. The ambulances lined up like cabs along the curb of Washington Square East.

At times, the young doctors and the policemen working with them among the dead thought they saw signs of life. "Then," said the *World,* "they would bring from under a mass of corpses a breathing woman. And as they carried her toward an ambulance, policemen, doctors and firemen fell back. Men covered their ears with their hands to shut out her cries."

The police could handle the dead; it was with the living they now had trouble. For the crowd continued to grow in size, its massed grief swaying it against the police lines which formed a circle half a mile in circumference around the Asch building.

Thousands of relatives and friends of Triangle's workers poured into the open spaces of Washington Square Park. There too came the curious, the disaster-lovers, the crowd-seekers. At the corner where Washington Place fronts on the Square, the pressure of the crowd became threatening.

"They pleaded, demanded and stormed to be let through. As time passed and more definite knowledge of the horror's proportion spread," the *World* reported, "the pressure on the police grew greater. The lines were reinforced but the frantic beating upon them outweighed every precaution of the authorities.

"Then someone in the Washington Square crowd cried agonizingly the name of a girl and rushed blindly at the police ranks. It started the huge crowd. The police were swept away. One thousand men and women had torn through the lines."

The policemen rallied, closed ranks quickly, and drove the crowd back out of the cleared area. But before they had done so, "men and women had rushed to the tarpaulin-covered mounds and knelt and prayed. . . . Here and there," the *World* continued, "a woman caught up a hat or a slipper or a fragment of a burned skirt, grasped it wildly and ran to a fireman or policeman, begging them to throw the light of their lanterns on the things.

"One of these women, looking at a poor, stained shoe, fell into a pool of water and was carried away struggling and utterly mad with grief as if she had surely recognized in it a daughter's possession."

The policemen prepared for a repetition of the break-through. It came at 6:45 P.M.

A portion of the huge throng, about five hundred at the same end of Washington Place, "pushed forward until the ropes went down and, shrieking, they swept over the police-men like a wave."

The movement was checked immediately. But it was only by force that the women were held back. "In more than one instance," the *World* added, "before the women gave way before them, the policemen were compelled to use their clubs."

Then there were no more breakthroughs. The crowd had been purged of its wildness. Sorrow settled down heavily on it, and the men now stood dumb and staring, the women with their hands to their mouths, sobbing, watching the panto-mime of the rescuers and the savers of life, lifting and carry-ing the dead.

It was done with great orderliness. Deputy Commissioner Driscoll, Inspectors Schmittberger and Daly, and Captain Henry of the Mercer Street station took charge.

On the east sidewalk of Greene Street, across from the Asch building, they spread a huge canvas of dark red. Working silently, in pairs, policemen carried about forty bodies of women, "charred and dripping" said the *Sun,* and laid them in rows on the canvas.

There they lay, waiting for coffins. And while they waited, the police covered them gently with tarpaulins and crossed back again to the sidewalk under the windows from which the dead had leaped and began to glean the cheap belongings the girls had clutched as they fell through space.

"There were leather handbags, broken combs, hair ribbons, some dimes and cents, parts of clothing," the *Sun* reported. In the gutter water, a policeman picked up what appeared to be a necklace but which under the electric light proved to be a rosary.

On the sidewalk, a russet shoe with all buttons gone, its heel half ripped off, evidence "that the girl who had worn it had caught her heel on a wire or projection when falling and it had been torn from her foot." In the street, a patent leather Oxford, its laces still tied in a knot and "hanging over its edge, a soiled garter." A fur-trimmed hat "with what had once been a gay red rose."

The coffins arrived. "Worming their way between the clutter of ambulances, mounted policemen, patrol wagons and throbbing fire engines," the *Sun* reported, "came men bearing rough brown coffins on their shoulders. The police had sent to the morgue for 75 or 100 coffins. But all they had was 65."

They began to pile the occupied coffins into the patrol wagons and ambulances. As soon as a carrier had its load, it moved off, its bell clanging. The silent crowd, pressed against the police lines, silently parted, opening a single lane for the passing dead or dying.

The city's hospitals filled with the near-dead. In the galloping dash to the hospitals, internes nursed their charges in the ambulances, applying splints and tourniquets, seeking to keep alive the small flicker of life but often finding it extinguished upon arrival.

More serious than the broken bones, said the *Sun,* were the cases of those who were badly burned about the face and body. They were quickly wrapped in oil-soaked bandages and taken to operating rooms. All the victims were suffering from shock, and this, the *Sun* added, "was just as liable to result seriously as was a fracture or a burn."

Shortly after six o'clock, teams of firemen spread through the building in search of victims.

On the second floor they found Maurice Samuelson of the cloak firm of Samuelson and Company. He had heard a commotion in the street and had gone to the front window of his office to see what was causing it. He opened the window. Without warning, a body shot by. Several more came rushing down in the next few seconds.

When the firemen found him, Samuelson was still standing at the window, immobilized by horror, frozen by fear. He whispered to them hoarsely that he could not move. Gently, the firemen led him to the street.

By 6:20, a team of firemen led by Chief Croker himself worked its way into the three top floors of the building. The woodwork on the tenth floor was still burning. Fireman Charles Lauph was one of the first to enter the ninth floor from the Washington Place stairs. He stumbled on bodies.

"I found two in a little crevice behind the dressing room; I found eleven bodies in front of the dressing room," said Fireman Lauph.

At the same time, Captain Ruch got to the ninth floor from the Greene Street side. He reached the Washington Place side of the shop, about six windows from the Greene Street side. Window frames were still burning.

"In my hurry, I stepped on something soft," Captain Ruch remembered. "I looked down and saw it was a body. I saw three or four others next to it."

Exploring the three floors, Chief Croker saw sights, "that utterly staggered him, that sent him, a man used to viewing horrors, back down into the street with quivering lips," the *World* reported. "In the drifting smoke, he had seen bodies burned to bare bones, skeletons bending over sewing machines."

Other firemen dug their way into the rear courtyard to which the Asch building's single fire escape descended.

The backyard was L-shaped. The smaller wing on the west side had three windows facing the New York University building. The longer portion of the yard, running the full length of the north wall of the Asch building, was 75 feet long and 25 feet wide. Of the eight Asch windows facing this part of the rear yard on each floor, the two center ones—the fourth and fifth windows—served as exits to the single fire escape.

A tenant in the Waverly Place building backing on the same courtyard said that iron shutters on these two windows, on every floor, had been closed for months. Triangle employees like Ethel Monick, for example, who had come to work for the firm in the fall, had therefore never seen the fire escape.

The balcony of the fire escape was 14 feet 6 inches long and 3 feet 6 inches wide. Each window facing it, when fully open, provided a clearance about 54 inches square. The ladder leading from floor to floor was 18 inches wide, cutting into the balconies 4 inches from their outer edge and 20 inches from the wall of the building.

The dark bottom of the courtyard was divided lengthwise by a cement wall. On top of the wall, which separated the Asch yard from the adjacent yards, was a 6-foot-high iron fence, topped in turn with 4-inch spikes.

The fire escape ended abruptly at the second floor. The remainder of the descent was by a drop ladder 12 feet 9¾ inches long. Hanging free from the second floor, the ladder ended 5 feet 9½ inches above the glass skylight of a ground-floor extension pushed out into the yard.

(A decade earlier, Building Department Inspector Miller had directed that Asch construction plans be modified so that the fire escape "lead down to something more substantial than a skylight." The architect had promised: "The fire escape will lead to the yard.")

The firemen found the skylight smashed. Its iron framework was twisted and broken. Pausing to look skyward as they gathered up the bodies from the dark pit of the yard, they saw the twisted structure of the fire escape, loosened from the side of the building, its top story gooseneck ladder to the roof torn away from the side of the building.

At each of the three Triangle floor levels, the iron window shutters, as Fireman Lauph later described them, were "sprung and warped." They had blocked the escape to life with almost the same finality as a locked door.

The *Evening Journal* attempted a reconstruction of the horror on the fire escape.

"A girl climbing out of the ninth floor window onto the fire escape," it reported, "passed down the stairs to the eighth floor safely and turned into the treadway. So far, she was alright. But in order to reach the steps leading down to the seventh floor, she must follow this little gangway across the windows to a similar platform-opening leading down to the seventh floor from the far side of the platform.

"The heavy sheet-iron shutters opening outward were in the way. Not only did they open outward until they completely blocked the gangway but they also buckled back in the direction of the wall so that they ran out in a V shape from hinges at the window to the railing of the treadway."

At each window, the shutters opened like two arms. In each case, one arm ended flush against the short side railing of the fire escape, extending at right angles to the building.

Abe Gordon, and a handful of others, seconds ahead of the crush that was to follow, had been able to squirm out of the ninth floor window facing directly on the ladder to the eighth floor balcony. Despite the heat and panic, they had somehow managed to bypass the shutters on the eighth floor.

But by the time Gordon reached the sixth-floor window through which he reentered the building, two terrible developments tipped the scale. The fire had driven a mass of panic-stricken women onto the fire escape. On both the eighth and ninth floor balconies the inner halves of the shutters, which were supposed to swing back flat to the face of the building in the space between the two windows, actually were locked outward.

"The shutters were held in place by a heavy iron rod,"

continued the *Evening Journal.* The rod should have been held fastened on the shutter by a hook. But this time it had not been hooked. Instead, it dropped down through the iron treads or cleats of the gangway and unless a girl got down on her knees and pulled out this heavy bar, the advance to the next steps would be impossible.

"The condition of the fire escape shows that the girls got down to the eighth floor and that the heavy bar holding the shutters was so firmly caught in the cleats of the treadway that they could go no further."

The fire escape grew heavy with its human load. At one of the ninth-floor windows some of the women and girls, finding their way blocked, turned and tried to go up the ladder to the tenth floor. But the fire blazed from the tenth-floor windows which opened on the packing and shipping room. There was no escape that way.

Those who came out of the other ninth-floor window had started down to the eighth floor only to be stopped in a long line blocked by the jammed shutter at the eighth-floor landing. On both floors, small groups struggled to get past the shutters in front of the open windows.

The screaming mass of people became immobilized. Then, out of all six Triangle windows fronting on the fire escape the flames roared forth. The heat and the weight of the struggling bodies bent the iron slats and railings, twisting them from the tenth floor down, crumpling them against the side of the building and dumping the flaming human load to the yard below.

A. D. Feldstein, manufacturer of hats and caps in the Waverly Place building which backed onto the courtyard, heard screaming and ran to his rear window.

"I hope I never again hear anything like it," he said. "It looked to me like a big pile of rubbish coming out of the windows, off the fire escape."

Down they came, crashing through the skylight and setting the room underneath afire. "As the fire-crazed victims were thrown by the collapse of the fire escape," the *Herald* added, "several struck on the sharp-tipped palings. The body of one woman was found with several iron spikes driven entirely through it."

Darkness had closed in. At 6:45 a crew of firemen finished rigging two block-and-tackle hoists from the roof near the corner of the building. Others, in the interior of the building, continued to gather the bodies.

A fire engine with a searchlight had been stationed on Greene Street and another on Washington Place, their rays busily searching the faces of the building.

On the upper floors, the firemen wrapped the bodies, sometimes two and three at a time; they had run out of nets and began using tarpaulins. The bundles of dead were carried to the gaping corner windows, where they were hooked onto the block and tackle.

They came slowly down on both sides of the building corner. "As each body was started downward," the *World* wrote, "the bright beam of a searchlight picked out its dark outline and followed it to the street. A fireman was on duty at each window it passed in its descent. He reached out and swung it clear of the sills. On the sidewalk a squad of policemen reached up as each body came to the end of its spotlighted journey."

The hellish performance was greeted only by the wailing of the bereaved men and women and children who had come to search for their own, which, said the *American,* made "a plaintive note above the hum and chug of the engines."

At about 8:15 P.M. Battalion Chief Worth and a group of firemen working on the ground floor of the burned building near the Greene Street entrance heard faint cries for help. Listening intently, they decided the cries were coming from below.

Carrying lanterns, they went down to the basement, where they splashed through water up to their hips. Boxes floated around them. Their lanterns cast only short rays into the darkness so they decided to use a device by which miners locate trapped companions.

They broke up into two parties and started in opposite directions to circle the basement. At the same time they set up a system of connecting shouts by which to take a bearing on the victim they sought.

They finally placed his cries as coming from the southwest corner of the building. They broke through three partitions and battered down an iron door to reach him.

He was in the bottom of the elevator shaft. At first, they saw only his head, directly over the cable drum on which the elevator cables were wound. Just above him was the floor of the elevator that had slipped down from the lobby level.

His name was Hyman Meshel, the *Times* reported. "Crazed by fright and blackened by soot, he was sitting helplessly on the elevator cable drum with his body immersed almost to the neck in water which was slowly rising in the basement."

The flesh on his hands had been torn from sliding down the elevator cable. His knuckles and forearms were full of glass splinters. His face was swollen and charred by the heat. "His eyes bulged from his head and he whimpered monotonously like a timid and spirit-broken animal," the *Times* added.

Several times Chief Worth shouted to him, "Get up! We've come to help you!"

Meshel made no reply. When they reached him, they carried him head high through the dark, flooded cellar and up to the street. He was rushed to St. Vincent's.

He had been in the rising water for four hours. In the hospital his wounds were treated and he was massaged in order to restore circulation to his paralyzed legs. Then he was able to tell how he had ended up nearly drowned in a blazing building.

He was working on the eighth floor when the fire started. In the excitement he had run to the Washington Place elevator. The door to the elevator was closed. Somehow, he beat in the glass upper portion of the shaft door and swung himself over the lower half. In the shaft, he could see the underside of the elevator above him. He grabbed one of the cables and began to descend hand over hand. But he grew faint and began to slip, tearing the flesh from his hands. He dropped to the bottom of the elevator shaft.

How long he lay there he did not know. But when he regained consciousness, he found he was lying in rising water. He could hear it splashing from a great height. Dazed, bruised, bleeding, he began to climb up the side of the elevator pit. But grease pots and wiping cloths on the pit embankment had caught fire, and it seemed to him that there were flames all around. He crept back into the elevator pit where he crouched in the water to escape the unbearable heat.

"As the water rose in the basement, Meshel began to fear that he would be drowned," the *Times* reported. "He climbed up on top of the cable drum and sat there with his back braced against the wall while the water crept slowly up to his neck. The cold so paralyzed him that he was unable to move, and the fear that after suffering so much he would be drowned made him semi-conscious."

The Edison Company strung rows of arc lights along Greene Street and Washington Place. Between eight and nine o'clock, they also strung lights through all the floors of the burned building. At 9:05 the lights in the building were suddenly turned on.

The shimmering glare lit up each gaping window. Huge, distorted shadows raced across the ceilings and walls visible through the window frames to the stunned crowd in the street.

Somewhere in the hollow of the building a burglar alarm began to ring—triggered into life by a broken wire. "It rang and rang," said the *Tribune*. "Nobody thought to stop it. Nobody thought to investigate it. It rang as a knell all through the night for the black, shapeless bundles being lowered from the windows."

As each body was received in the street, it was carefully searched for jewelry or other personal effects that might be useful later for identifying the body. The small things were put in an envelope on which was written the same number inscribed on the coffin into which the body was placed.

By eight o'clock, sixty bodies had been lowered and tagged.

Death wagons that had already deposited their first loads began to return for a second, bringing back the emptied coffins which would be used again.

It had become clear to Charities Commissioner Drummond and his first deputy, Frank Goodwin, that the city morgue was too small to hold so many dead. He directed his men to convert the covered pier at the foot of East Twenty-sixth Street, at the far end of which were the offices of his department, into a temporary morgue.

Deputy Police Commissioner Driscoll ordered a temporary police station to be set up on the pier and put Captain O'Connor of the East Twenty-second Street station house in charge. Forty men from various precincts were directed to report there at once. Of these, twenty-five were assigned to form a line across Twenty-sixth Street at First Avenue.

The dreary journeys of the death wagons from the Asch building to the temporary morgue went on and on. Most of them drove up Broadway to Fourteenth Street, into Fourth Avenue to Twenty-third Street, then turned east to First Avenue and on to Twenty-sixth Street. Each wagon, its bell clanging through the crowded streets, carried three or four bodies and several policemen.

At the pier the coffins were lifted from the wagons under glaring arc lights that had last been used for a similar purpose in 1904 when the *Slocum* sank in the East River. "In cases where the bodies were not burned too seriously," said the *Sun*, "they were wrapped in shrouds and laid out on the floor of the pier with heads to the wall on the two sides of the building. The bodies that were badly burned remained in the coffins."

This was the reason for the shortage of coffins. By ten o'clock, thirty-three badly burned bodies meant that not enough coffins were being sent back to Washington Place; the dead lay there waiting.

Commissioner Drummond sent the Charities Department ferryboat, *The Bronx*, to Blackwell's Island in the middle of the East River to get the coffins that were stored in the carpenter's shop of the Metropolitan Hospital. The steamer returned with two hundred at about 10:30 P.M. and "from that time on there was no delay for lack of coffins," the *Sun* added.

There was also a shortage of manpower on the pier. Commissioner Drummond called on William C. Yorke, superintendent of the Municipal Lodging House on East Twenty-fifth Street, for as many men as he could muster to handle the bodies in the emergency.

At about ten o'clock, Superintendent Yorke, heading a double line of 24 derelicts, came marching through the entrance of the temporary morgue. Throughout the long night, this group of the city's disinherited lifted the bodies, cared for the dead.

At the Asch building the mocking burglar alarm continued to ring. At 11:15 another body was lowered from the ninth floor. When it reached the street a helmeted fireman leaned out of the ninth-floor window and shouted to those below: "That's all on the upper floors!"

At 11:30 o'clock, Chief Croker wearily told the reporters: "My men and I have gone through every floor, every room in the building. We have gone through the basement, we have gone through the airshaft at the rear of the building. Every body has been removed."

It was, the *World* summed up, "the most appalling horror since the Slocum disaster and the Iroquois Theatre fire in Chicago. Men and women, boys and girls were of the dead that littered the street: That is actually the condition—the streets were littered with bodies. At intervals throughout the night, the very horror of their task overcame the most experienced policemen and morgue attaches. The crews were completely changed no less than three times."

The Mercer Street station house of the 8th Police Precinct was only three blocks from the Asch building. It became the first focal point of the frenzy of those who sought but could not find their loved ones.

They had been kept at a distance while doctors and firemen worked on the bodies on the sidewalk. Under police escort, they were taken to the Mercer Street station house. Before the overwhelming proportions of the disaster were realized, the station house had been designated as a headquarters, and the first injured were brought there.

Then an order was issued that no more injured were to be brought to the station house. A line of policemen covered the entrance and inquirers were told to proceed to the morgue.

But from six o'clock on, crowds of women kept coming into the precinct building. Even after the doors were shut "the crowd of crying mothers and sisters was lined up by twos and when admitted, were asked to give the names of those they sought and then to go to the morgue," the *Sun* reported.

Mention of the morgue made the women hysterical. A policeman opened the door of the section room to come out. "In a wink," added the *Sun*, "the crowd had burst into the patrolmen's sitting room, vainly looking for bodies. The women screamed and beat their foreheads on the desk rail."

One of those who stormed the precinct house was Gussie Horowitz, a tucker who had escaped from the eighth floor. Back and forth, she elbowed her way through the crowd, calling her brother's name. Rose Katz grabbed her arm.

"Your brother Morris is all right," she shouted to Gussie above the din. "He came down from the eighth floor with me. He was a little hurt. He sent me to look for you!"

The blotter of the Mercer Street station house confirms the fact that while the bodies were being transported to the morgue on the pier, the small objects which had linked them to life were brought to the police station.

Some of the first of these were turned in by Patrolman Meehan who had come galloping down Washington Place even before the arrival of the first fire engine. The day book entry on Meehan ends on a practical note that comes as a relief in all the horror: "At 4:55, while putting an unknown woman who had been injured in the fire into an automobile, he caught his coat and tore the tail of same."

By 9 P.M. the miscellany of scraps of lost lives entered on the precinct record included: "One lady's handbag containing rosary beads, elevated railroad ticket, small pin with picture, pocket knife, one small purse containing $1.68 in cash, handkerchiefs, a small mirror, pair of gloves, two thimbles, a

Spanish comb, one yellow metal ring, five keys, one fancy glove button."

Under "property found by Fire Chief Ed Croker" the following is listed: "One lady's handbag containing one gent's watch case number of movement 6418593 and a $1 bill, one half-dozen postal cards, a button hook, a man's photo, a man's garter, a razor strap."

Then: "Patrolman Edward Clark brought to this station one portion of limb and hair of human being found at fire. Sent to morgue in patrol wagon."

Toward midnight, many of the stricken moved dazedly in the circuit from Washington Place, to Mercer Street Station, to the morgue on Twenty-sixth Street, and then back to Washington Place. The early morning scene at the Mercer Street station house was heart-rending, said the *Times*.

Those who came there returned to it as a last resort, knowing that it contained none of the dead, yet hoping desperately that they might now recognize a ring or watch or purse and return with the knowledge of its number to identify a loved one at the morgue.

"Many turned away from the morgue and the Mercer Street station house to take positions again in the lines surrounding the factory building when they found there was no trace of their lost ones among the bodies in the morgue and no trinket at the station house which would enable them to claim some body burned beyond recognition," the *Times* wrote. "Broadway, at midnight, in the vicinity of Washington Place, was thronged with women walking up and down and wringing their hands while calling the names of their kinfolk whom they had lost."

The vigil continued through the night.

At dawn, the area was flooded by tens of thousands of the bereaved and the curious. The former were easy to recognize for they were "wan and weary, having tramped from morgue to hospital to police station and everywhere getting the same answer, 'Only God knows.' At dawn they found their way back to the blackened building that had proven a mausoleum for so many young lives."

A soft-spoken priest wandered through the crowd. "This sorrow is economic as well as sentimental," he told a reporter. When he turned from the priest, the same reporter saw an old woman, standing away from the crowd, take a small phial from her pocketbook. It had a death's head on it. He sprang forward and wrenched it from her fingers, at the same time calling a policeman who stood nearby.

"This is the sixth that tried that trick," the officer told the reporter for the *American.*

As the early morning sun began to rise, a Salvation Army trio standing in the Square opposite Washington Place struck up "Nearer, My God, to Thee." Few sang. Few even wept; there were no tears left.

A block away, on Waverly Place, a dozen desperate relatives stormed the police line and rushed forward toward a huge plate-glass window on the street level of the building that shared the backyard with the Asch structure. Their plan was to crash through the glass, rush to the roof and lower themselves down to the Asch ruins. The police drove them back, "begging them with tears in their eyes to be patient," said the *American.*

Early in the morning, after pumping engines had drawn the water out of the flooded basement, the search for the missing was resumed in that area of the building. Two more bodies were discovered hanging over the steampipes behind the boiler.

They were the bodies of two young women who had crashed through the Greene Street sidewalk. Both bodies were mangled beyond recognition. On the right arm of one was a gold bracelet with a heart-shaped locket initialed "L. T." The other still wore small diamond-set earrings.

Now, in the morning, Chief Croker was on the ninth floor for the fourth time since the fire. He spotted a mouse stirring in a corner. It was half drowned.

He picked it up and stroked it. Then he put it in his pocket, telling the two firemen with him he would take the creature home.

"It's alive," he said. "At least, it's alive."

8. DAY

A death by violence, and painful wounds,
Are to our neighbor given.
—CANTO XI:34

STARTING early Sunday morning, great throngs of men and women, many with children, descended from all parts of the city upon the area around the Asch building. Thousands began to form into a slowly moving parade around the city blocks closest to the burned building.

In Washington Square, others settled down for the day, selecting a place to stand, a tree against which to lean, a railing on which to sit, remaining there, immovable, even after it was announced that no more bodies could be found in the building.

Shortly before ten o'clock, Inspector Max Schmittberger drove up to where Washington Place branches off from the Square and conferred with Deputy Police Commissioner Driscoll. The throng was growing larger by the minute, and the after-church and postprandial crowds were still to come.

A decision was made and the police whistles shrilled as the patrolmen firmly and steadily pushed back the crowd on all sides. When they were finished they had cleared the area bounded by Waverly Place on the north, Mercer Street on the east, West Fourth Street on the south, and the edge of Washington Square on the west.

The reporter for the *Tribune* studied the crowd, moving against the stream and at times picking a group or an individual for closer scrutiny. Occasionally, where four or five were huddled together, he heard a sob of despair and the replying murmur of consolation. Others stood conferring and pointing, trying to read from the distance the marks and signs of the disaster on the face of the building.

But there were also those on whom the impact of the tragedy was lost—"irresponsible young fellows with giggling girls on their arms who seemed to regard the whole affair as a festival."

The privilege accorded the press enabled him to move within the area that had been cleared. For several moments he stood in the middle of the crossing of Washington Place and Greene Street where he could not hear the throng but could only see it down all the four directions—"as in a battle square formation."

It seemed strangely silent from his vantage point. Silently, firemen and policemen hurried through their cleanup assignments. Even orders and signals from man to man were given in low voices as if in respect for those who had been carried away so short a time ago.

But when he walked down the middle of Washington Place toward the police line at the Square, the color and sound of the crowd came alive again. He spied in the Square a bright-colored hat, a small boy perched in one of the old trees. He passed through the police line and moving through the crowd crossed the street and entered the Square. "Every place of vantage from which any angle of the building could be seen was occupied."

He continued his walk westward toward Washington Arch at the foot of Fifth Avenue. Had he not known of the toll of death the fire had taken his "journey through these people in the park would have been most misleading. The general atmosphere was that of a country circus or of a Vanity Fair."

At the intersection of two footpaths stood a vendor of hot jellied apples, leaning on his small pushcart, calling his wares in a strident voice. "If those who passed close by him did not heed his voice he had recourse to his small bell which he tinkled energetically as he extolled the virtues of his stock."

When the *Tribune* reporter came to the Washington Arch he saw a procession of "unwieldy Fifth Avenue stage buses," heard the "resonant blowing of their horns and the rumble of their heavy wheels." Atop the open buses, he noted, were gaily dressed men and women. "The hats of the women would form a gaudy bit of coloring against the dull green of the bus, like flowers in a bit of greensward, and the high silk hats of the men would add a touch of smartness to the effect."

Now he turned and walked again in the direction of Waverly Place. Even the ringing of the conductor's bell atop the buses jarred him.

At Waverly Place he fell in with the crowd on its mournful march. He found it easy to distinguish the curious from the stricken. Those for whom some part of their world had ended "walked slowly—as slowly, that is, as the police would allow them, with a curious halting step. Many of them kept their hands folded across their breasts—the women—and there was piteous appeal in their eyes."

At Greene Street he left the parade, walking down the block again to Washington Place, finding again the center where the two streets crossed. "In the center of this square of sorrowing humanity, rising like a giant tombstone, the blackened and scarred building reared skyward."

And around and around and around it, in a kind of macabre Maypole procession, the mourners moved. At the corner of Greene Street and Waverly Place they could see that corner of the building where the fire had started. Then they would cross narrow Greene Street, and the next building cut them off from the reporter's view. He had made a mental mark of three of them—the old woman in the full skirt with the apron over it, the apron she had forgotten to remove as she started to run, yesterday; the old man with the short beard, the bowler hat, and a jacket with a sleeve that had ripped loose at the back shoulder; the boy young enough to be wearing short pants and ribbed black stockings.

In less than five minutes he spotted them again, this time slowly crossing Washington Place at the Mercer Street side of the cleared area. Again, they had slowed their pace, walking forward but with their eyes fixed on the building at this new angle from which they could see the place where the girls had plunged to their deaths in groups.

Early in the afternoon the man from the *Tribune* and a reporter from the *Times* were allowed to tour the Asch building.

The reporter for the *Times* found that on the top three floors, "the walls, floors and ceilings were intact as were the pillars which support the ceiling. Only the woodwork was burned away."

On the tenth floor, in the area of the Triangle offices, he noticed a heavy safe and reported that officials took this as proof of the strong construction of the building for "had the flames been able to weaken the floors the safe would have gone crashing down through the building."

Not a stick of furniture, except the iron structural parts of the machines, had been able to resist the flames. "The flames must have swept the great square rooms from side to side, leaving not one little corner in which refuge could be found."

He inventoried the smaller links with life. "On the eighth and ninth floors were found more than two dozen rings, at least 14 of which the police say were engagement rings. Almost a bushel of pocketbooks and handbags were taken out of the debris."

The *Tribune* reporter found the roof of the Asch building undamaged and concluded that there would have been no loss of life if all had been able to reach it. Two skylights had been shattered by the heat from below. There were a few small pools of water on the cement and gravel surface of the roof. He saw the marks and scratches on the side of the adjoining Greene Street building where many had climbed to safety.

Across the smaller narrow court on the west side of the building he saw the iron shutters of the New York University Law Library "hanging twisted and torn from their hinges while the glass of the windows was gone and the furniture inside seemed scorched and scarred by the terrible heat of the flames that had not penetrated the school building."

He examined the Washington Place staircase, noting that "the walls of the stairway shaft were unscorched and unmarked by any sign of fire except immediately around the doors where, as the fire burned its way through the doors, the whitened walls had blackened for two or three feet only. The fire had not entered the stairway shaft."

On this staircase he carefully examined the hoses and reported that in no instance had a hose been attached to the standpipe system. On every floor, the hose still hung next to the standpipe "and the couplings had never been put in place for use while the doors were closed between the hose and those whom it might have saved."

He checked the elevators on the Washington Place side. On all three floors, the doors to the right-hand elevator shaft were closed. The doors to the left-hand elevator shaft were closed on the eighth and tenth floor but open on the ninth "and the elevator at the bottom of the shaft showed for what terrible purpose it had been torn open."

The *Tribune* reporter returned to the tenth floor and went down on the Greene Street side. He found that "like the staircase on the Washington Place side this one, too, was wide enough for only one person to pass at a time. To pass another person, one had to wait on one of the wider steps which occurred at every sixth step where the stairs turned in the narrow shaft."

On each of the three floors he found the freight elevators closed behind heavy iron doors. "On the ninth floor there was a heavy iron bar before them. There was no sign that the doors had been opened on any of the floors during the time of the fire."

As the darkness deepened late on that Sunday afternoon, the crowd began to thin. Tomorrow would be a workday. By ten o'clock the last stragglers had left and only the firemen working in the building and the small group of policemen guarding it remained. The bereaved had carried their sorrow home or to the morgue.

Deputy Police Commissioner Driscoll estimated that during

the day after the fire more than fifty thousand persons had visited the scene.

But early Monday morning there was a crowd again. Many had postponed a view of the burned building, planning to view it on the way to work. During the noon hour, when hundreds poured out of the nearby factories, there were nearly as many as on the day before.

In the afternoon, a small group of well-dressed young women, said the *Herald*, began to gather under the canopy of the New York University building on the Washington Place side.

The police had allowed them to pass through the lines. They were Triangle employees.

Each newcomer was warmly greeted, as if she had just returned from a long journey. In whispers and with tears "this one or that related for the hundredth time how narrow had been her escape from the flames and how poor Sadie or Esther or Mamie might have escaped too if she had only done thus and so."

Not all of the reunions were happy ones. The *Herald* told of the young woman who said her name was Jennie Detora of Brooklyn who came through the police line and ran to a group of girls, asking if any had seen or heard of her cousin Alberta.

"One of the girls remembered having seen Alberta in a perilous situation on the ninth floor. While she was telling her story, Jennie Detora threw up her hands and collapsed on the sidewalk in a swoon. An ambulance took her to St. Vincent's."

And all the while, hawkers moved through the crowds. These enterprising fakers found eager customers. They had quickly assembled a stock of penny rings, glass jewelry, cuff links, and other trinkets. These they packaged in small match boxes and envelopes.

Up and down the streets they went, crying:

"Here they are! Get 'em while they last! Souvenirs of the big fire! Get a dead girl's earrings! Get a ring from the finger of a dead girl!"

9. MORGUE

Who is this that without death
Goes through the kingdom of the dead?
—CANTO VIII:84

BEHIND the closed doors of the Twenty-sixth Street pier, the thirty regular morgue attendants and Superintendent Yorke's supplementary force of indigents and panhandlers prepared the dead to meet the living. The cavernous pier, said the *World*, "is the stage of all the abject tragedies of the city. Its floor is crossed by the insane, the paupers, the mean criminals, the helpless and the diseased on their way to the cold stone public institutions, charitable and penal, set on the islands in the river."

At the time of the *Slocum* disaster, the Twenty-sixth Street block ending at the river had been given a new name—Misery Lane. On Saturday, it had begun to fill with a growing crowd even before the first ambulances arrived. By seven o'clock, there were two thousand people in Misery Lane. A cordon of policemen marched across the width of the street from the pier, sweeping everyone back to First Avenue. Twenty-five of them, with sticks drawn, remained at that point to keep the crowd from pouring down the block and into the pier.

As the first dead wagons swung around the First Avenue corner, they were followed by weeping, screaming men and women who were stopped by the cordon of police. The horse-drawn ambulances and patrol wagons passed down the block and through the huge pier doors. When the wagons had deposited their freight, they backed out slowly and returned for more bodies.

The derelicts and the doctors worked among the dead, the latter in the hope that someone might have survived. Coroners Holtzhauser and Hellenstein arrived at seven o'clock to join Coroner's Physician Weston in supervising the work.

At the First Avenue police line the pressure of the crowd increased. "Mothers and wives had run frantically through the street in front of the carriers, pulling their hair from their heads and calling the names of their dear ones," the *Times* wrote. When they were stopped at First Avenue and Twenty-sixth Street, their sobs and shrieks grew louder.

As each new wagon arrived, police with drawn clubs fought to open a path. But the frantic people closed against the rear of the wagons as soon as they passed, and hands reached out for the blankets and tarpaulins covering the dead.

Once a cover was torn off, revealing underneath two bodies somehow holding each other. "Everywhere burst anguished cries for sister, mother and wife, a dozen pet names in Italian and Yiddish rising in shrill agony above the deeper moan of the throng," said the *Times*.

Inside the morgue, confusion arose over the numbering of the dead and their possessions. The police had used small colored tags provided by the Charities Department, and these were found to include some duplicate numbers. Moreover, the coroner's men had renumbered some of the bodies and the jewelry envelopes.

Inspector Richard Walsh settled the problem by deciding to renumber all the bodies in one continuous count. Valuables were to be turned over to Lieutenant Sullivan for safekeeping. If a body could be identified, the lid was to be placed on the coffin and the entire thing was to be moved to one side. Until this procedure was complete, the crowd would be kept out.

A reporter for the *Times* who had been watching these preparations came out of the pier gate on his way to a telephone. He found himself besieged.

He wrote: "A hundred faces were turned up to him imploringly and a hundred anguished voices begged of him tidings of those within.

"Had he seen a little girl with black hair and dark complexion? Had he seen a tall thin man with stooped shoulders? Could he describe any one of the many he had seen in there?"

Inside, in the deepening darkness, small groups of policemen moved from box to box. One held pencil and pad; the others carried spot lanterns. They would stoop over an open box, and one of them would pull back the sheet covering the body.

"Box 112," he droned to the one with the pencil and pad. Then, lifting pieces of clothing out of the box into the light of the lantern: "Female. Black shoes. Black stockings. Part of a brown skirt. White petticoat."

On the first time around, an estimated $5,000 in cash was taken for safekeeping by the police from the bodies on the pier.

In the case of bodies which seemed easy to identify, the police removed valuables in properly numbered envelopes. But "where the bodies were so utterly incinerated as to make recognition impossible by other than bits of jewelry, the finger rings, necklaces and earrings were not removed," said the *World*. "Policemen were especially assigned to guard these coffins."

Outside the clamor to get in rose in waves. To relieve the pressure on First Avenue, the police allowed two lines, four and five abreast, to form down each side of Misery Lane. A rumor spread that the police would open the pier gates at midnight. It threw the people into a "wild hysteria of almost joy," the *Times* wrote. "Several women had to be taken to Bellevue for treatment, laughing and crying and struggling all the way."

In an effort to placate the crowd and ease the coming ordeal in the morgue, a policeman walked the length of the waiting lines; at the sound of his voice the people strained toward him, listening:

"Who seeks a girl with a ring bearing the initials G. S.?"

A shriek cuts the air—an old woman breaks out of the line, staggers forward.

"Who seeks a girl whose pay envelope bears the name of Kaplan?"

By 11:30 P.M. the police in the morgue had "processed" one hundred and thirty-six bodies. "Smiling Dick" Walsh, his face damp where he had wiped away tears, added that of this total fifty-six were burned or crushed beyond physical recognition. Every stitch of clothing on dozens of bodies had been burned off. The body of one girl was headless.

Outside, the thousands stood shivering; the night was cold. Their anguished cries filled the air. Inside, the police, the coroner's doctors, the superintendent's little army of disinherited had finished their work.

It was midnight.

The great iron gates swung open.

Now everything grew silent. In the crowd, only a moment ago so anxious to enter, no one stirred.

A policeman called out, "Come on! Come on! It's all ready now!"

Slowly, an old woman with a shawl over her head started into the pier. All others remained still, their eyes on her. But when she had taken some twenty paces, suddenly the whole crowd moved. Inside the pier, policemen stationed at the coffins began to swing their lanterns over the faces of the dead, for the sputtering arc lights seemed to cast more shadows than illumination.

The little shawled woman moved down the center aisle, leading the crowd. A third of the way down the line she stopped at the coffin numbered 15. She fell to her knees, cried out. The first identification had been made.

Her scream stopped the crowd. Some turned as if to run away. But others pressed forward. Minutes later a dozen had found what they had sought, caught in the spotlight of a policeman's lantern.

Shivering and weeping, the waiting people were admitted

in groups of twenty, each group accompanied by several policemen who almost immediately had their hands full.

The bodies lay in long rows, covered by white sheets, their heads propped on boards so that those passing could more easily make identifications. At first glimpse of the ghastly array, many fainted into the arms of policemen and nurses. Superintendent Yorke's panhandlers brought them hot coffee.

As soon as an identification was made an officer put the lid on the coffin, tacked onto it a small yellow card on which he wrote the name of the victim, and sent the relatives, with a corresponding slip and name, to the temporary office of the coroner at the side of the pier. Permits to remove the identified bodies were then issued to them.

At one time during that weird morning, the arc lights failed. In the darkness, half-crazed mourners ran about, falling over coffins. Swinging lanterns, as if floating in the air, came to them, carried by nurses or officers ready to help and to pacify. After four minutes, the lights came on again. Six women and two men were found to have fainted.

Small groups would cluster around several of the boxes, seeking for familiar features. For some, recognition was accepted silently with a sinking to the floor or fainting. Many men broke into hoarse sobs, terrible to hear. But with some it was heralded by piercing shrieks, hysterical lamentations in Yiddish or Italian that brought the nurses running and then struggling to pry despairing arms loose from the coffin.

The experienced morgue attendants had arranged the bodies so that those in the worst condition were farthest from the doors. In this way it was thought the hardships for friends and family would be minimized.

Desperate parents moving deeper into the long pier somehow sensed the logic of the arrangement. As they neared the end of the double row of dead, many women became hysterical. The police, thinking that an identification had been made, would start to close the coffin nearest the shrieking woman only to find "they had been misled by the woman's despair," said the *Times*.

Dominick Leone of 444 East Thirteenth Street didn't

scream. He came looking for three cousins and a niece, and he walked on tiptoe among the dead, as if fearful of waking them.

But Clara Nussbaum straightened up from the coffin over which she had bent for a moment and ran screaming to the edge of the pier. Captain O'Connor reached her as she was climbing over the barrier below which ran the river and dragged her back. She had recognized her daughter Sadie, eighteen years old. The lower part of Sadie's body had been almost totally destroyed by the flames.

Commissioner Drummond ordered all the pier openings to be boarded up to prevent any more suicide attempts.

The fainting and the hysterical were taken to the northeast corner of the pier, where Dr. Louis W. Schultz, General Medical Superintendent for the Department of Charities, had set up an emergency hospital. Working with him were the Misses Blair, Robertson, Hamilton, Cook, and Donovan, students from the New York Training School for Nurses; they were all under twenty years old.

They were "a legion in themselves," said the *Tribune*. "In a scene of such horror and terror, men were glad to have those quiet, gentle-mannered women around. They soothed the terrified, took the arms of the poor, creeping, old mothers."

The paper told how one nurse stood with her arm around a youngster who had brought a picture of her sister for identification.

"Aren't you frightened?" the child asked as she shuddered and buried her face in the clean white linen apron.

"Frightened?" came the reply. "No, my little one. Why should we be frightened? The poor things can't hurt anyone."

Joseph Miale of 135 Sullivan Street made the circuit of the dead three times.

"I am looking for my sister, Bettina," he told Captain O'Connor. "She is not here."

But as he turned to leave the pier, he pointed to one of the bodies and said to O'Connor that it seemed to be the same height as his sister. The Captain drew a ring from the finger of the body. He showed it to Miale who staggered backward, crying, "That is her ring."

In the first hours after midnight no one came to claim No. 138. She was easy to identify. She was about thirty-seven years old and weighed about 130 pounds. The two policemen who had searched for some means of identification on the body were amazed to find $852 in bills.

The money was pinned inside the left stocking and held fast to the leg by a band of cheese cloth. The two men had first seen a lump inside the woman's stocking. It was watersoaked and hard, and when they first saw it they thought it was a physical deformity in the leg.

But when they cut away the stocking and the cloth, they found a roll of one, five, and ten-dollar bills. In one of her pockets was a pay envelope with $10. The name on the envelope was Mrs. Rosen.

Her treasure was intact. But Mr. and Mrs. Morris Bierman, of 8 Rivington Street, complained that the body of their daughter, Gussie, had been stripped of three rings, a watch and chain, and earrings. Gussie kept her cash in her shoe and one shoe was missing. The family insisted that the loss was no accident. Gussie's rings had fitted tightly on her fingers; she habitually wrapped the chain of the pendant watch twice around her neck. The remaining shoe was proof, they said, that the missing one must have been removed by force.

By sunrise, forty-three bodies had been identified. The official recognition accorded the dead was scant. The cards tacked to the boxes read: Anna Cohen, Box No. 31; Celia Eisenberg, Box No. 56; Julia Aberstein, Box No. 32. On the lid of one coffin a strong hand had written with white chalk: "Becky Kessler, call for tomorrow."

During the early morning, the line of those seeking admission to the pier lengthened, till it stretched from the pier to First Avenue then down from Twenty-sixth Street to Twenty-second Street.

In the afternoon, the character of the crowd began to

change. The poorly dressed men and women who had stood in line for hours for the chance to identify a relative were being replaced by more fashionably attired persons.

There were, said the *Sun,* frock-coated young men carrying canes who laughed and chatted with well-dressed girls as the line jostled slowly toward the pier entrance and young couples "who had read the morning papers before starting for their Sunday stroll and who wished to see the dead out of curiosity."

Five young women, stopped at the entrance by a policeman, brushed him aside and ran into the pier. They were hustled out by other officers.

Once an old woman emerged from the morgue supported by two other women. Her wrinkled face was fixed with a weird grimace, the tears ran down her cheeks, she cried aloud and threw her arms in the air, said the *Tribune.* "And a great boisterous crowd followed her down the avenue, pointing and peering until the police drove them back."

The police also had to cope with fashionable women who drove up to the police line in their cars and asked to be let through. When told they would have to take their place at the end of the line, "they turned up their noses and ordered their chauffeurs to drive off."

A policeman at the gate, in midafternoon, counted one hundred persons a minute entering the morgue—six thousand every hour. Yet the crowd didn't seem to diminish.

A squad of officers went down the line in an effort to weed out those who obviously had no business there. But suppose even one of these could identify one of the bodies? The result of the weeding was a recognition of the fact that many who had come out of the morgue had returned to the line for a second visit. But the police had the satisfaction, as the *World* put it, "of yanking out and putting to flight not less than forty known pickpockets come for the purpose of nipping jewelry from the seared throats and charred fingers, of thumbing stealthily burned clothing in the hope of finding a purse."

At five o'clock, Deputy Police Commissioner Driscoll arrived. He studied the endless line and said to the officers around him: "Good God! Do these people imagine that this is the Eden Musee? This doesn't go on another minute!"

He ordered his captains and lieutenants to remove anyone who could not immediately give the name and the description of the relation or friend he was seeking.

In ten minutes the crowd in Misery Lane had been reduced from thousands to several hundred. Even these were closely scrutinized by Nurse Mary Gray, an assistant to Dr. Healy of the Charities Department, who was stationed at the entrance to the pier to question all who sought admission.

She had a wonderful searching eye, said the *World,* and many found themselves stammering foolishly when she questioned them. She turned them back "with sweetness and light," but cleared the way quickly for those "in whose eyes she could read genuine grief and suffering."

A big lieutenant standing nearby smiled. "She's worth twenty cops," he said.

Inside the morgue the cataloguing of the dead continued.

Rosie Solomon of 84 Christie Street took her position in the line leading to the morgue early Sunday morning, but it was one o'clock in the afternoon before she entered. Once inside, she made the circuit of coffins, looking only at the hands of those who lay in them.

When she reached Box No. 34 she stopped, her gaze fixed on a ring she seemed to recognize. She asked the attendant if there was also a watch on the body. There was. He opened it and she looked inside, seeing her own picture.

Rosie Solomon sank in a faint to the floor. She had found her fiancé, Joseph Wilson, by means of the ring and the watch she had given him. They were to have been married in June.

For Serafino Maltese, a young typesetter, grief was triply compounded. First he identified the body of his sister Lucia, aged twenty. Then he found the body of his sister, Rosalie, aged fourteen. At that point he fainted. As soon as he was revived, he began his sad march again, this time to search for the body of his mother, Catherine, aged thirty-eight.

Box 74 held the body of a woman whose regular features,

said the *Sun,* "were without scratch or stain." When she had worked in the factory the day before she must have looked very much as she did now in the pine box. But no one among the thousands who went by recognized her.

Outside the pier, since early in the morning, a line of black hearses had waited. Even before dawn, a common sight became that of two men carrying a rude coffin to one of the hearses, followed by their group of wailing women.

All day long, the *Herald* reported, "the somber equipages rattled over the rickety pavement, followed by the awe-stricken glances of the spectators." There was much bantering among the undertakers who "chaffed each other good naturedly. For one day, at least, there was business enough for all."

Between midnight and seven o'clock Sunday evening, close to 200,000 had come to the area around the temporary morgue. More than half had walked among the coffins.

Beginning on Sunday, the city's newspapers published all through the week the list of the unidentified dead.

At midnight, thirty-three identified bodies, including two brought from New York Hospital, and fifty-five still nameless were moved to the regular morgue, 100 feet south of the pier. The work was done by Superintendent Yorke's corps of derelicts, who carried the coffins through the light rain that had started to fall. One box contained only a skull and some charred bones.

Shortly after midnight another group of men was called up from the lodging house. These entered the empty pier, carrying brooms and mops and pails with which they cleaned away the stains and prepared to fumigate and disinfect the Charities Department pier.

The fifty-five bodies were set up head to head in two rows in the center of the rotunda floor of the morgue. Around them, the dull gray walls were lined with the square doors of vaults reserved for the normal flow of the city's unknown and violently dead.

Early in the morning, the police had some difficulty with

the crowd near the exit of the morgue. A number of times hysterical women rushed toward a small group carrying out a coffin and threatened to seize it. The police believed these were relatives who had not yet found the one for whom they searched and were fearful that now that the number of bodies was diminishing, some one might claim theirs in error.

Though Monday was a workday, the crowd was only slightly smaller than the day before. Nurse Gray, backed by six burly policemen, held her post at the entrance to the morgue. Among those she stared down through her pince-nez glasses were some young men who claimed to be medical students but could not prove it.

At one time during the day word spread through the waiting crowd that immediate admission could be gained by asking for "the girl in the blue skirt." The *Sun* said about five hundred did ask but were told "there were no blue skirts around any of the dead."

But in the course of the long afternoon, many did find their own. Ignatzia Bellotta's father came in from Hoboken to look for the body of his sixteen-year-old daughter. When he found her, he knew her by the heel of her shoe. The *Times* reported, "He had taken her shoe to be repaired and the shoemaker had put in a plate whose peculiar construction he recognized."

Five times that day, two men came to look at the body in coffin 138. They were certain she was Mrs. Julia Rosen of 78 Clinton Street. Commissioner Driscoll, remembering the $852 found in Mrs. Rosen's stocking, insisted that a close relative of the victim be produced before any claims would be met.

They produced her. At 4:30 in the afternoon, the two men returned, leading fifteen-year-old Esther Rosen, the woman's daughter. She leaned over the box and touched the head of the dead woman. "It's mamma's hair. I braided it for her. I know—I know."

Esther and two younger brothers had waited at home all day Saturday and Sunday for their mother and brother, both of whom worked at Triangle. Then the police knocked on the door. Esther didn't know the meaning of the word "bank" but explained that the family had come to America four years ago, the father had died, and the mother always feared to leave the family's savings at home.

One mystery had been solved; another remained.

Who was the beautiful woman in the coffin numbered 74? Her hair and her face were undamaged. When leaping to death she had apparently folded her arms. Officials were baffled by the inability of any of the Triangle survivors to recognize her. Then Mrs. L. Bongartz, a matron on the municipal ferryboat running between the Charities pier and Randall's Island, said she recognized the woman as a mother who often made the trip to visit her child in the hospital. Miss May Dunphy, superintendent of nurses on the Island, was directed by Deputy Commissicner Goodwin to come to the morgue with any other nurses who might help in the effort to identify the woman.

Dominick Leone, tiptoeing hat in hand around the bodies, had found his two cousins, Nicolina Nicolose and Antonina Colletti, while they were still on the Charities pier. He came to the morgue Monday evening with other members of the family to look for his niece.

A hailstorm broke with sudden ferocity as they entered the rotunda. The glass panes forming the ceiling had been opened. The arc light hanging from the center of the dome began to swing wildly in the wind and the rain and hail fell on the faces of the dead.

Attendants ran to close the roof panes. Leone and his relatives ignored the entire disturbance, quietly moving from box to box until one of the group hesitated. Where they stopped, Dominick leaned over the box, brushed aside the long hair partly covering the face. One of the women in the group cried out. The search had been successful.

Not all of Triangle's dead were in the morgue.

Two who had leaped for their lives lay in the New York Hospital fighting to live.

 Daisy Lopez Fitze of 11 Charlton Street didn't have to

work. Her husband had recently returned to Switzerland with their small savings to buy an inn. He had left his wife enough to live on until he would send for her. But Daisy was determined to save every penny so she went to work at Triangle.

She died. And they kept word of her death from her friend in the next ward who had also jumped. The doctors said Freda Velakowsky of 679 East Twelfth Street still had a chance.

At 9 P.M., Monday, twenty-eight unidentified bodies remained in the morgue.

On the third day after the fire, the crowd began to form in Misery Lane shortly before dawn. It was the smallest crowd since the catastrophe. Before noon, half of the sixty patrolmen and mounted policemen on duty were sent back to their regular assignments.

They had hardly passed from sight when about one thousand factory girls, dressed mainly in black, marched into Twenty-sixth Street at First Avenue. Under the impression that there was to be a public funeral of the unidentified that afternoon, they had come to pay their final respects to fellow workers. They were dispersed by the police.

Benny Castello went to the shoe store on Houston Street and purchased a pair of button shoes like the single pair his sister had bought in this same store. In the morgue he identified her by matching the shoes he carried with those in the pine box.

Hope of identifying the well-preserved, Italian-looking woman in Box No. 74 evaporated. No one from Randall's Island could recognize her. Commissioner Drummond ordered that pictures be taken of her and that they be reproduced in the city's Italian-language newspapers.

In New York Hospital, Freda Velakowsky lost her fight.

The police moved the jewelry and other personal possessions of the victims to the East Thirty-fifth Street Police Station.

Twenty unidentified bodies remained in the morgue.

Four days after the fire, pretty Mary Leventhal was still among the missing. Many of the workers, with whom she had been a favorite, had come to look for her. But it was Joseph Flecher, who worked in the Triangle office, who remembered he had recently sent her to a dentist friend of his.

He came to the morgue with his friend, Dr. J. Zaharia, who recognized her by a gold cap he had recently set.

But Sophie Salemi's mother recognized her by the darn in her stocking at the knee of her left leg. "I mended it only the day before the fire," her mother said. "That is my Sophie."

Policeman Peter Purfield, who had already been complimented by Commissioner Drummond for his resourcefulness in the identification work, closely examined a shoe taken from the body of the Italian-looking woman in No. 74 box. He found a baby's sock in the heel.

Wednesday night, sixteen bodies remained to be identified.

On Thursday, Sarah Kupla, sixteen, died at St. Vincent's Hospital. She had leaped from the ninth floor and had never regained consciousness.

The aged mother of Antonina Colletti complained at the coroner's office that the body of her daughter, just returned to her, had been robbed of $1,600 sewn into the hem of her skirt. It represented six years of her earnings. The coroner had no record of the money.

Salvatore Maltese told the police that his wife, Catherine, had not been seen since the fire. He had been to the morgue every day. He could not identify her.

By evening of Thursday, the band of unidentified had been reduced to fourteen.

Little Esther Rosen had identified her mother on Monday. On Friday, she identified her brother, Israel, seventeen, by a signet ring.

Joseph Levin of Newark, New Jersey, had a sister who boarded with a family on Delancy Street. They called him on Thursday to tell him that she had not been home or heard from since the fire. He came to New York Friday morning.

His sister's name was Jennie. She was nineteen. He found her easily. She was the pretty one, the untouched one, the Italian-looking one in Box 74.

Ten victims remained to be identified.

Joseph Brenman identified his beloved sister, Sarah—Surka, as he had called her during the ninth-floor turmoil. Anna Ardito was identified. Gussie Rosenfeld was identified.

One week after the fire, of the 146 dead, only 7 remained without names.

PART TWO

10. GUILT

Whereby the heavens were scorched.
—CANTO XVII:108

AT 11:20 P.M. on the day of the fire, Coroner Holtzhauser and Deputy Fire Commissioner Arthur J. O'Keefe stood among the last bodies on the Greene Street sidewalk waiting to be transported to the morgue.

"The Fire Department will have to meet a terrible responsibility," said Holtzhauser.

"And so will some other departments," was O'Keefe's quick reply. "Fire Commissioner Waldo reported this place to the Building Department within the past three months as a building unsafe to use as a factory."

The great debate had begun. Who was to blame for the tragedy?

Politicians and bureau officials, anticipating public wrath, searched for the language and logic with which to justify themselves and escape blame.

On Sunday, scores of city officials inspected the Asch building. Starting Monday, the city's newspapers set up a clamor for the quick determination of responsibility. Political axes were sharpened. Four separate investigations were launched: by the Coroner's Office, the District Attorney, Fire Commissioner Waldo, and Acting Building Department Superintendent Albert Ludwig.

The *Tribune* began to carry on its front page a standing

box on the fire. One of the first of these read: "If any individual is guilty, by negligence or greed, he should be punished. Chief Croker warned New York after the Newark disaster. His warning was ignored."

The *Evening Journal,* on the other hand, disdained such restraint and instead daily published on its front page the drawing of a gallows with the caption: "This Ought To Fit Somebody; Who Is He?"

From highest to lowest, public officials expressed themselves as appalled by the tragedy. With the same unanimity, all denied individual fault or departmental responsibility.

Among the first to insist he was appalled but also without responsibility was Governor John A. Dix, who declared: "I am appalled by the terrible disaster in New York. I find that I am powerless to take the initiative in an inquiry. My advices are that the city authorities alone are the ones to deal with this problem. They possess all the jurisdictions needed. I must refrain from discussing the question of placing Borough President McAneny on trial in the absence of charges formally made. It is hard to believe that such a thing could happen in this day and age."

W. W. Walling had been, until March 1, the First Deputy State Labor Commissioner and had authorized an investigation of the Triangle factory because his department had received a complaint. He explained:

"The State Labor Department has no supervision in the matter of fire escapes in Manhattan. At the time of the shirtwaist strike, we investigated strikers' complaints that too many persons were working on a floor in the factory of the Triangle Shirtwaist Company. The complaint was unwarranted. Under the law, each worker must have 250 cubic feet of air space and this the Triangle employees had."

The First Deputy State Labor Commissioner had stared at the killer and had not seen it, for at Asch, the proud modern fireproof building, the ceilings were sufficiently high to supply the air space required by law—while the crowding on the floor of the shop could be that much greater, and more dangerous.

State Labor Commissioner John Williams also "deeply deplored the tragedy." In the next breath he added, "I am glad that my conscience does not have to bear the burden of responsibility for it."

The Commissioner then pointed out that the courts had held the State Labor Department had no jurisdiction in the matter of fire escapes. In City of New York versus Sailors' Snug Harbor, the Court of Appeals had declared in 1903: "We think it was the intention of the legislature to leave with the City of New York jurisdiction for the proper officers of that city over the subject of fire escapes upon factories."

The state was unable to act. "When our inspector finds there is no fire escape, we send to the Superintendent of Buildings of the proper borough a special blank calling his attention to the matter with a request to advise us what has been done about it," the State Commissioner of Labor added.

In the case of Triangle, the proper borough, of course, was Manhattan, and the proper Superintendent of Buildings was Rudolph P. Miller—the same Miller who a decade earlier had found fault with the Asch building plans. He was absent from the city when the tragedy occurred. But he could have found no stauncher defender than Borough President George McAneny, who stated:

"While it is true that I am responsible for the Department of Buildings, I am sure that no person expects that I shall personally go out and inspect each building to see that all provisions of the building code are complied with."

Miller had been appointed superintendent of the department because the Borough President "believed him the best qualified man in New York City for the post. . . . He has worked for the city steadily for fourteen months, early and late, with no rest whatever. He has been at his desk up to nine and ten o'clock, night after night, working over problems."

Unfortunately, the dynamo of public service was out of the country. McAneny continued: "He was invited by the government to inspect new methods of steel and concrete work in the Panama Canal and I urged him to go as we in

New York City are vitally interested in the most advanced methods in the use of reinforced concrete."

But others were not as ready to compliment Superintendent Miller or the department he headed. District Attorney Charles S. Whitman was another in the great army of civil servants who "was appalled by what I saw when I arrived at the fire." He had ordered "an immediate and rigid" investigation for the purpose of "determining whether or not the Building Department had complied with the law."

Coroner Holtzhauser went further and insisted that "the matter rests entirely with the Building Department. The Civil Service is rotten. Any drygoods clerk or anyone else can pass the examination and they are appointed inspectors. Hardly any one of them has any knowledge of buildings. What should be done is to appoint men who know something about the construction of buildings and how they should be equipped."

But to Acting Building Superintendent Ludwig, "the inadequate force of the Building Department is not our only trouble." The Department, he pointed out, had no power to police and "we must enforce all our rulings through the civil courts. When we bring an action there is invariably a long fight. The record will show the owner is usually the victor."

Ludwig also emphasized the fact that the Department had only forty-seven inspectors with approximately fifty thousand buildings to inspect in Manhattan alone. In February, 13,603 buildings were listed as dangerous by the Fire Department, but the Building Department's inspectors had been able to visit only 2,051.

With more than $100,000,000 worth of new construction annually in the area of its jurisdiction, it is impossible, Ludwig added, "for us, with our limited number of inspectors, to make regular examinations of the buildings that have once been passed as filling the terms of the law. The best we can do is to look into specific complaints that are brought to us from outside sources."

The Acting Superintendent admitted, "It is quite true that the Asch building conformed to the law at the time that it was built and that there are many changes that would be demanded at the present time. It is also true that the department has the power to order any changes it sees fit to make a building conform to the law as it is at present."

Nevertheless, he insisted that "we do not hear of violations of the law in old buildings unless they are particularly called to our attention." Furthermore, "It would often work a great hardship on the owners of the building to require extensive changes. This is especially true of fire escapes."

After viewing the twisted fire escape Coroner Holtzhauser had come out of the lobby of the Asch building "sobbing like a child," the *Times* reported. Angrily he declared:

"Only one little fire escape! I shall proceed against the Building Department along with the others. They are as guilty as any. They haven't been insistent enough and these poor girls who were carried up in the elevator to work in the morning, came down in the evening at the end of a fireman's rope."

Acting Superintendent Ludwig again admitted that the "fire escape was undoubtedly of the size and style ordered by the department at the time the building was put up but it would not be passed today."

But this was no fault of Superintendent Miller, said Ira H. Woolson, spokesman for the National Board of Fire Underwriters. "There is no question that the emergency exits from the building were foolishly inadequate," Woolson admitted. But there was no evidence to show that "this condition was attributable to the neglect or inefficiency of Mr. Miller." On the contrary, said Woolson, Miller could neither make laws nor control such matters. "He must confine his efforts to the limits of the city statutes as he finds them."

Borough President McAneny was even more insistent in his defense of Miller. It was "outrageously unfair" to attempt to place the blame on the Building Superintendent. The Borough President declared that he wished to state, "—and I cannot put it too strongly—that the present officers of the department have no responsibility in the matter whatever."

Again, he pointed out that the judgment as to whether the

exits in a building are proper and sufficient is made at the time building plans are filed. In the case of the Asch building, this was in 1900, and the plans "were acceptable then as complying fully with the law."

This was, in fact, not true. Miller himself, when he was only an inspector, had found fault with the Asch plans and had called for a structural change in the fire escape. And even as the law stood in 1900, the building should have had three staircases.

McAneny acknowledged that the city code gave the Superintendent power to direct that fire escapes and "other means of egress" be added to buildings of the Asch type "where they are found lacking." But the Building Department, as far as McAneny said he could learn, had "never attempted to go back to the original approval of plans or to order changes except in particular cases in which there was evidence of failure to comply with the law."

The Building Department relied on the Fire Department to inspect for such evidence. No report about the Asch building had been forwarded by the Fire Department.

Speaking for the Fire Department, Chief Croker replied that he had long predicted such a loss of life as had occurred in the Asch building "unless fire escapes are put on all buildings in which there are a large number of persons." He stressed the fact that "at the present time there is no law compelling the construction of such fire escapes," and clearly located the responsibility for ordering them: "The matter is entirely within the discretion of the Building Department Superintendent."

The Chief also said irritably, "There wasn't a fire escape anywhere fronting on the street by which these unfortunate girls could escape. I have been arguing, complaining and grumbling about this very thing for a long time. But every time I raised the point some of these architects and 'City Beautiful' people would pop up and declare that to place trappings of iron and steel upon the front of buildings would destroy the beauty of the city. My answer to their argument is this fire."

One of "these architects" was Francis H. Kimball, who readily admitted that Chief Croker was probably hitting at him. Kimball declared that his own objection to fire escapes on the fronts was not because "they would be unsightly, for a fire escape can be planned along good lines and made a really artistic adjunct to any building." He preferred, he said, the type of fire-staircase required by law in Philadelphia. This he described as an entirely separate stairway leading to the street, which has no doors opening into the interior of the building and leads on each floor to outside iron balconies.

"Outside fire escapes," Kimball noted, "while they may save a few lives, usually fail in high buildings. Who would want to go down a fire escape on the outside of a twenty-story building? Chief Croker isn't afraid to go anywhere but we are not all like him. I climb about a good deal but I feel uneasy on a fire escape up in the air. And when you get frightened women and have flames and smoke pouring out of lower windows, outside fire escapes are not much protection."

District Attorney Whitman also expressed sharp concern with the fact that the stairway doors opened inward. He said he was under the "impression" that this was a violation of the law and that inward-opening doors had been the cause of the terrible loss of life in the Chicago Iroquois fire in 1903.

Both Acting Superintendent Ludwig and Architect Julius Franke, who had worked on the original plans for the Asch building, gave the answer. Ludwig cited the "where practicable" condition modifying the directive that factory doors be made to open outward.

But Architect Franke was even stronger in his insistence that "the building was put up in strict compliance with the law." Concerning the stairs, he said the law prescribed no "particular" width so that it bore no relation to either the square footage of each floor or the height of the building. Secondly, said Franke, it was considered that the stairs in the Asch building would be sufficient "to accommodate what it was supposed would be the traffic of the building."

"When it was put up," Franke added, "it was only intended for a loft and business building," with apparently no thought, according to the architect, that it would become a tower of factories.

Acting Superintendent Ludwig had this in mind when he declared that Building Department Commissioner Brady had approved the plans in 1900 and that the completed building also was favorably considered by the Fire Department, the Department of Water Supply, Gas and Electricity, the Health Department, "and all other departments charged with inspection and approval. Since that time there has never been one complaint made against the Asch building.

"Even the drop ladder at the bottom of the fire escape was in compliance with the law. The worst feature was that the escape ended in an enclosed court. There is nothing in the law to prevent this," Ludwig declared.

Ten years earlier, Inspector Miller had considered that ladder arrangement to be faulty. Architect Franke's explanation, however, indicated that then the danger had not been so great. "When the building was put up," he said, "there was an opening from the court in which the fire escape ended through an alley about 15 feet wide. Since then, the old buildings in that block have been torn down and new ones put up that have enclosed the court on all sides."

Time joined the other villainies.

"Even so, the Asch building is the only one on the block with a fire escape," said Franke.

"If it had only been called to the attention of the Department, we would have ordered a change," Ludwig asserted.

He had insisted several times that there is "no law governing the dimensions of the stairways." They needed only to be "of a size we consider sufficient."

But Fire Commissioner Rhinelander Waldo declared unequivocally that "in the opinion of the Fire Department, the means of exit from the Asch building were not good and sufficient as the law requires."

"The employees were woefully ignorant of the layout of the building," Fire Marshal William L. Beers pointed out. "Loft and factory buildings where people speaking different languages are employed should have placards in their languages telling them how to get out in case of fire. There should be fire drills. Quick exit is essential. The heat in the huge room where these girls were trapped became quickly so intense that they dropped where they stood, as flowers might wither under the same influence."

"Those responsible for buildings," Chief Croker bitterly remarked, "include the Tenement House Department, the Factory Inspection Department, the Building Department, the Health Department, the Department of Water Supply, Gas and Electricity and the Police Department to see that the orders of the other five departments are carried out. Yet, the Fire Department doesn't have a word to say about fire escapes or fire exits."

Who was responsible?

The conscience of an outraged city found voice in Lillian D. Wald. As a member of the Joint Board of Sanitary Control she had helped draw up the report written by Dr. Price, warning of the danger in the shops. She said:

"The conditions as they now exist are hideous and damnable. Our investigations have shown that there are hundreds of buildings which invite disaster just as much as did the Asch structure. The crux of the situation is that there is no direct responsibility. Divided, always divided! The responsibility rests nowhere!"

11. HELP

. . . dripping is with tears.
—CANTO XIV:113

THE TRIANGLE tragedy set off a chain reaction of misery. In the city, death struck grief into the hearts of more than one hundred families. It also left them without breadwinners in a foreign land. The injured needed help; the dependent were stunned.

Outside the city, in far-distant lands, the slow but welcome flow of help from a relative in America suddenly stopped in scores of families. The horror echoed in homes in Russia, Austria, Palestine, Jamaica, Hungary, Roumania, and England.

The city opened its heart and its pocketbook. Public-spirited citizens quickly launched a massive relief effort through the Red Cross Emergency Relief Committee of the Charity Organization Society. Its chairman was Robert W. de Forest and its treasurer Jacob H. Schiff, the Wall Street investment banker.

The day after the fire, de Forest called on Mayor William J. Gaynor and urged him to issue an immediate appeal for contributions. This the Mayor did, starting the drive with a personal contribution of one hundred dollars. In his message to the citizens of New York he declared:

"The appalling loss of life and personal injuries call for larger measures of relief than our charitable societies can be expected to meet from their ordinary resources. I urge all citizens to give for this purpose by sending their contributions either directly to Jacob H. Schiff or to me for remittance to him."

The emergency committee rallied notables from New York society. It included on its roster Otto T. Bannard, Cleveland H. Dodge, Mrs. William K. Draper, Lee K. Frankel, Mrs. John M. Glenn, Lloyd C. Griscom, Thomas M. Mulry, Leopold Plaut, Mrs. William B. Rice, and Dr. Antonio Stella. The secretary of the committee was Edward T. Devine.

At eight-thirty Monday morning, the committee opened an office in Room 11 of the Metropolitan Life Insurance Building arcade at 1 Madison Avenue. The insurance company donated the space and the furniture. Mr. Devine was in charge.

By ten o'clock a staff of home visitors, clerks, and stenographers was on hand, recruited with the aid of the United Hebrew Charities, the Association for Improving the Condition of the Poor, the Society of St. Vincent de Paul, and the Charity Organization Society. The Police Department provided the corps of visitors with lists of the dead and the injured.

By noon, the trained visitors were making the rounds of the bereaved homes. By Tuesday morning they were completing a revised card record of the victims; by Wednesday evening, every family on the police lists had been visited.

While the city-wide relief effort was getting under way, the garment workers launched their own aid campaign, in coordination with the drive of the emergency committee. On Sunday afternoon, Local 25 of the International Ladies' Garment Workers' Union held a meeting of its membership at Clinton Hall in the heart of the lower East Side.

It issued an appeal addressed not only to other units of organized labor but also to the general public:

"Everybody knows that the victims of this terrible catastrophe were poor working people whose families are either left destitute or can ill afford the cost of sickness or death. The sole support of the family in many cases has been swept away. This is an occasion when everyone can give his share and feel assured of assisting a worthy cause."

The union committee was especially helpful in guiding the Red Cross workers to the stricken families and in overcoming the language barrier. A joint committee that, in addition to Local 25, included the Workmen's Circle, a Jewish fraternal organization; the Women's Trade Union League; and the *Jewish Daily Forward* cut across class lines and received the endorsement of such social leaders as Mrs. O. H. P. Belmont, Mrs. James Lees Laidlaw, Mary E. Richmond, Mrs. Henry Ollesheimer, and Mrs. Mary K. Simkhovitch. It received $100 contributions from Mrs. J. P. Morgan, Anne Morgan, and Mrs. Walter Lewisohn. Working with the committee were the Reverend John Haynes Holmes and the Reverend William H. Melish.

When the Red Cross emergency committee opened its doors on Monday morning, it had $5,000 ready for immediate distribution. But nobody came to ask for aid. On Tuesday morning, four relatives of victims turned up. They came not to ask for help but because they had been told to "report themselves." The teams who visited the homes of the victims had found that the families, in most cases, were too shocked by grief to consider their own economic plight.

"We went into the East Side to look for our people," says Rose Schneiderman. "Our workers in the Women's Trade Union League took the volunteers from the Red Cross and together we went to find those who in this moment of great sorrow had become oblivious to their own needs.

"We found them. You could find them by the flowers of mourning nailed to the doors of tenements. You could find them by the wailing in the streets of relatives and friends gathered for the funerals. But sometimes you climbed floor after floor up an old tenement, went down the long, dark hall, knocked on the door and after it was opened found them sitting there—a father and his children or an old mother who had lost her daughter—sitting there silent, crushed."

According to the Red Cross, the families affected by the

tragedy were, for the most part, recent Italian and Jewish immigrants, largely dependent on the earnings of their girls and women working in the seasonal soft-goods industries.

These families had never before received charitable assistance; they did not seek it now. The Red Cross described them as being, in the main, naturally self-reliant. As proof, it cited the fact that "applications for assistance were received from only about half the employees who were in the fire and from very few who had not been physically injured although information about the fund and its great size was widely spread through the papers in their own languages." In a number of cases, the committee continued, "relief has been declined on the ground that the family resources were sufficient. The committee has respected and honored such self-reliance."

Meanwhile, money poured in from rich and poor alike. Checks arrived from businessmen with notes of sympathy; coins were wrapped in letters scrawled by children.

The *Times* printed "as received" a letter containing $10 written "in the uncertain hand of a child," whose name was Morris Butler. It read:

"Dear Mr. Editor: i went down town with my daddy yesterday to see that terrible fire where all the littel girls jumped out of high windows. My littel cousin Beatrice and i are sending you five dollars a piece from our savings bank to help them out of trubbel please give it to the right one to use it for somebody whose littel girl jumped out of a window i wouldent like to jump out of a high window myself."

Andrew Carnegie sent $5,000 to the Mayor's office.

Movie producer William Fox announced that receipts on Monday, Tuesday, and Wednesday at the New York Theatre, Broadway and Forty-fifth Street, would be turned over to the relief fund.

Marcus Loew offered the fund a day's receipts from his vaudeville theaters on Seventh Avenue, 149th Street and Third Avenue, the Lincoln Square, the Yorkville, the Circle, the Plaza Music Hall; and the Columbia, Bijou, and Liberty Theatres in Brooklyn.

At the Hippodrome, 500 employees raffled off a gold watch at fifty cents a chance and turned the proceeds over to the fund.

The Salvation Army set up coin pots at many places in the midtown part of the city.

Dan Morgan, manager of Valentine "Knockout" Brown, the lightweight boxer, said he and several managers of fight clubs intended to hold a series of benefits for the families of the fire victims.

Louis Roughburg's newsstand stood at the corner of Fifth Avenue and Eighteenth Street. On Monday he nailed across the top of the stand a sign painted in heavy black letters: "Will Give Today's Receipts to the Washington Place Fire Relief Fund."

The depth to which the public had been stirred could be measured by the action taken by Leopold Hallinger, a wealthy real estate owner. He announced he would turn over his two houses at 322 and 324 East Houston Street for three months of free occupancy to families which had lost their wage earners in the fire. After that they could remain for monthly rents ranging from thirteen to sixteen dollars.

Through its secretary, H. E. Adelman, the Hebrew Free Burial Society warned the Jewish public against giving money to "schnorrers," professional beggars going from house to house, pretending to gather contributions for the burial of fire victims.

"We are not asking a cent," said Mr. Adelman. "We'll stand it all alone. Of course, we need contributions badly but we are not sending around for them."

The society, made up of five thousand of the East Side's poor, had buried sixteen victims by late Monday. It announced it would provide in its Staten Island cemetery for Italian victims as well as Jewish because, said Mr. Adelman, "there is at present no Italian organization to take the place the society fills among the East Side Jews."

New York's great theatrical community, with traditional speed and generosity, prepared to help. On Sunday, Giulio Gatti-Casazza, director of the Metropolitan Opera Company, announced that the noted opera house was available free for benefit performances. The Musicians' Union and the Stage Hands' Union immediately offered to service benefit shows without charge.

Charles H. Burnham, president of the Theatrical Managers' Association, undertook to arrange a program. More than three hundred performers offered their services. Three program committees were named: Gatti-Casazza, John Brown, and William Hammerstein for opera; Winthrop Ames, Alf Hayman, William and Sam H. Harris for drama; Percy Williams, Arthur Hammerstein, Marcus Loew, and E. D. Miner for vaudeville. The date of the benefit performance was set for Tuesday, April 11.

Not to be outdone, the rival independent theatrical managers were marshaled by Lee and J. J. Shubert for their own benefit performance at the Winter Garden on Friday, April 7, which began at one o'clock in the afternoon and ended at six. Everyone—actors, stage hands, ushers—donated their services. Al Jolson was master of ceremonies. The benefit performance at the Metropolitan Opera House the following week was an even greater success. With George M. Cohan in charge of the acts, almost $10,000 was raised. With the proceeds of the Winter Garden show and other benefits, the New York stage had raised close to $15,000.

In the days following the fire, the headquarters of Waistmakers' Local 25, ILGWU, on Clinton Street remained draped in black from roof to the ground floor, and every day it was besieged by bereaved relatives. In the first two days after the fire, the union's committee provided for the burial of eleven girls. Under the plan worked out by the Joint Relief Committee organized by the union with William Mailly as secretary and Morris Hillquit as treasurer, the larger Red Cross committee referred cases in which the victim had been or still was a member of the union to this group for initial care.

The Red Cross was facing its first large-scale disaster operation in New York. It immediately set a policy aimed not so much at reimbursing financial losses as such but rather at restoring, as quickly as possible, the accustomed standard of living or preventing a serious lowering of that standard. It was also decided, for the first time, to reach across the world to help the families of victims of a local disaster.

The committee was able to do these things because of the size of the fund. The unprecedented liberality of contributions by the public, the committee felt, reflected a "passionate desire to do whatever remained in our power to compensate for the horrible event."

The grand total contributed was $120,000, a huge sum by contemporary standards. Of this, $103,899.38 was in the Red Cross Emergency Fund; the remainder was in the union's joint committee fund.

Far from setting a scale to compensate for a lost life, the Red Cross committee gave individual consideration to each case. To expedite this procedure a special conference of experienced social workers was organized, which included Mrs. Glenn, Dr. Frankel, and Mr. Mulry of the emergency committee; Mr. Adelman of the Hebrew Free Burial Society; John A. Kingsbury of the Association for Improving the Condition of the Poor; William I. Nichols of the Brooklyn Bureau of Charities; W. Frank Persons of the Charity Organization Society; William D. Waldman of the United Hebrew Charities; and William Mailly and Elizabeth Dutcher of the union's joint committee.

The committee soon discovered that in a number of cases in which it tried to "obviate the lowering of the standard of living, it has been found even easier to facilitate an actual improvement; to grant a lump sum that sets a father up in business, for example." In all cases, the committee exercised heart-warming compassion.

In the emergency committee's Case No. 130, the widow of a machine operator was left with two small children. She was given $1,000 in May to enable her to buy a small store in Yonkers. By November, she was doing so well that she had engaged domestic help in order to be able to give more time to "building the business." Even so, the committee set aside

1. The Asch building housing the Triangle Shirtwaist Company on the eighth, ninth, and tenth floors.

2. Tons of water were poured into the Asch building long after the fire was brought under control.

3. After the fire was brought under control, bodies were laid out and tagged on the sidewalk across the street from the Asch building.

4. Three officers carrying personal effects of those killed in the fire.

5. Friends and relatives identifying bodies in Pier Morgue after the fire.

6. Crowds gather outside the Asch building.

7. A room in the Asch building after the fire.

8. Joseph Zito, the elevator operator who performed heroic work during the fire.

9. Coroner and jury questioning employees.

10. Thousands line the street waiting for the funeral procession.

The Worthless Fire Escape and the Death Trap Below It

[illegible] VIEW OF THE BROKEN FIRE ESCAPE THAT LED MANY TO DEATH BY [illegible] [illegible] The [illegible] shows the [illegible] from the sixth and tenth floors. The frail iron balcony and [illegible] would accommodate only one person at a time. As many [illegible] gave way and they plunged to a sunken areaway which [illegible] The iron doors at the top, when shoved open by the fleeing girls, [illegible] cut off escape to the roof.

THE FATAL PIT BENEATH THE FIRE ESCAPE FROM WHICH TWENTY FIVE BODIES WERE TAKEN. [illegible] light through which bodies dropped to the third floor. B—Height reached by water in flooded [illegible] Iron picket fence on which two girls were killed. D—Steel barred basement windows.

11. A description of the disaster in a local newspaper.

12. The unions mourn.

13. One of hundreds of families in mourning.

14. Despite earlier safety warnings, the doors were kept locked. (From the New York *Evening Journal*, March 31, 1911.)

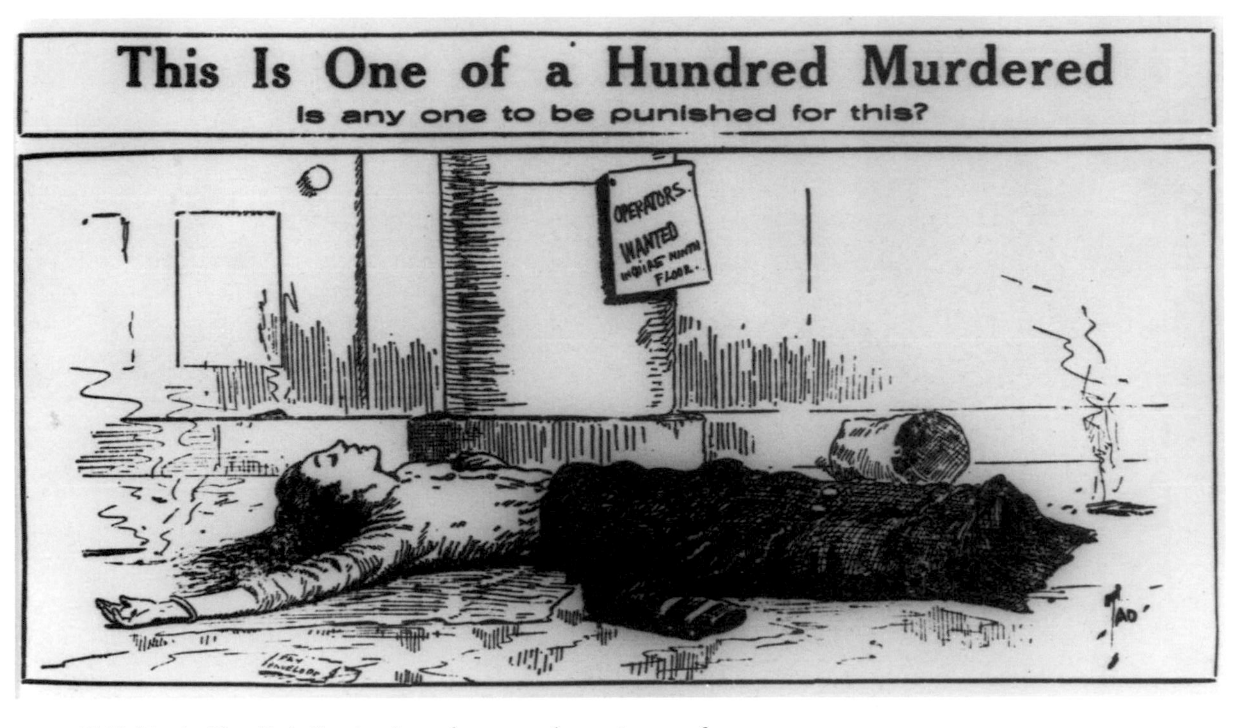

15. Tad in the New York *Evening Journal* expresses the sentiments of many.

16. A commentary on safety inspections.

a $3,000 trust fund for the two children.

Similarly, in Case No. 7, a widow was left penniless. But her two children, fourteen and twelve years old, were stranded with grandparents in Russia. The emergency committee made an award of $250 with which the widow bought a share of ownership in a stationery store. Seven months later she was able to send for her children.

But in Case No. 18, the attempt to help in this way failed. The widow, with an eleven-month-old baby, could speak no English. The report describes her as remaining in an hysterically anxious frame of mind for several weeks and as going from agency to agency for advice. Finally, she, too, decided to buy a stationery store. But her health was poor. The emergency fund financed a summer stay in the country. In November, though she was still unable to work, the fund continued its aid.

The Red Cross handled a total of 166 cases. In 94 there had been one or more deaths; in 72 there had been no deaths. A total of $81,126.16 was disbursed for relief. Individual family aid ranged from $10 to $1,000 in families where there had been no deaths and from $50 to $5,167.20 where one or more lives had been lost.

The committee understood that it was dealing with lives that had been wrenched by a horror. If it happened in some cases, "that money given for business has been used for current living expenses, we do not feel that this necessarily proves that the relief should not have been given.

"A chance has at any rate been given the family—heretofore capable of managing its own affairs successfully. It is not impossible that the failure—on its own responsibility—has been as valuable an experience as carefully guarded and guided success would have been. There is such a thing as pauperizing by too much advice and guidance."

The tragedy struck deep.

The wife of one of the two heroic elevator drivers who had stayed with his car to the end suffered a miscarriage on hearing of the fire. After that he himself suffered a decline in health. Their three-year-old child sickened. By April, they had been given a total of $400 but thereafter asked for no more. In October, husband and wife were still ailing.

In Case No. 89, the mother of a sixteen-year-old girl killed in the fire "was seriously affected by the shock and for a long time," says the Red Cross report, "it was impossible to rouse her from her depression.

"The neurologist who examined her thought there was grave danger of suicide. She could not be induced to go to a hospital. An Italian woman who lived in the same house voluntarily assumed all responsibility for her: prepared meals for her and did not leave her alone a moment for many weeks.

"The mother's mental condition slowly improved and she went back to work. When she began to get better she told about an illegitimate child, a baby girl, who was in an institution, and she became more and more determined to have this child with her. This was discouraged on account of her condition.

"But late in August she brought the child home. At first this seemed to have a bad effect on her mental condition. She called the child by the name of the dead girl and seemed even more depressed. Then she began to improve."

Both the Red Cross and the union committee remembered that the fire had occurred on the eve of important holidays. Sums ranging from $25 to $50 were given to needy Jewish and Italian families for Passover and Easter expenses.

In a number of cases it was clear that the most effective aid that could be given to an individual survivor would be a period of rest or convalescence in the country. The Association for Improving the Condition of the Poor arranged for a number of these cases to be accommodated at its Hartsdale home. The Solomon and Betty Loeb Memorial Home gave them priority consideration. Other institutions made similar offers.

But it was difficult to persuade the girls to leave their homes. Only twelve of the Italian victims could be induced to go. The approach of Passover increased the reluctance of the Jewish girls. Finally, only three accepted the generous offer of the Loeb home.

Arrangements to take care of survivors were made in eighty cases, and for these a total of $61,571.05 was disbursed. Of this amount, $44,672.15 was for relatives in the United States and $16,898.90 for relatives in other lands. In some of these cases, dependence was slight, but in others it had been "absolute and inevitable," the Red Cross declared.

The arithmetic of disaster was shattering. Several families were left helpless by the loss of a daughter upon whom they had relied for financial support and much more. Four men left widows with children. One young victim had been the sole support of his widowed mother and invalid middle-aged sister. In another family, two sisters were killed, one of them a widow with five children.

In five other instances, two sisters lost their lives; in another family two brothers. In the Rosen family, the death of their mother and older brother left three children without a natural relative. The Maltese family was cut in half by the death of the mother and two daughters; the father and his two sons, one nineteen and the other five years old, were left in utter despair and bewilderment.

In three of the eighty families, there were dependents both in the United States and elsewhere. One of these was the family of Daisy Lopez Fitze, who had, for two days, survived her leap to the sidewalk from the ninth floor. Aid was sent to her father in Jamaica. In Switzerland, where he had gone to start a small business in which his wife was soon to join him, Daisy Fitze's husband mourned her but never sent word to the Red Cross.

In forty cases there were dependents in the United States only; thirty-seven involved people in other countries. In seven instances the Red Cross committee helped dependents return to Europe and these included four Italian families. But a youngster left stranded in Russia by the death of his sister in the New World was brought to the United States to join relatives in New York.

To small towns scattered throughout Europe, messages came telling of children who had been burned or who had fallen to their death. While they had lived, there had come from them a small stream of the Golden Land's surplus, even the promise of future reunion in America. Both were ended.

Of the thirty-seven cases involving dependents abroad, only four were Italians, the committee noted, and pointed out that families in Italy were less likely to send daughters to the United States to start a new life than were the Jewish families in Russia and other east European countries where they faced prejudice as well as poverty.

The task of tracing the overseas families was undertaken by the Italian Consul General, the Jewish Colonization Bureau of Paris, and American Consuls General in Moscow, Vienna, and Bucharest.

Some who lost their lives had been recent newcomers to America who had used borrowed money and family savings to pay passage fare.

In one remote Russian village, a Jewish blacksmith who had borrowed money to send his oldest daughter to the New World was puzzled by a letter and 400 roubles ($206.20) from America. In the letter, he learned she was dead.

In another Russian city a father, six months before the fire, had grown tired of trying to support five children and a wife on the rouble a day he earned in a match factory. So he borrowed 100 roubles and sent his sixteen-year-old daughter on her way along a relay of relatives stretching across Europe to America. When she died she had been in the United States less than three months. Not a single rouble had as yet come back from her. Her steamship ticket had not been fully paid for. The Red Cross committee paid the balance due on the ticket, and sent 500 roubles to the father.

Some, at first, required no financial aid. For instance, there was Case No. 50. Two brothers had died. Their parents lived on a farm in Austria, purchased with $500 the boys had sent from America. The parents were self-supporting. The boys had planned a visit home for Passover. In October, the Red Cross learned that the death of her two sons had driven the mother insane, that as a result the father faced destitution. A grant of 2,500 kronen ($506.25) was dispatched.

Some didn't know where to turn. This is the committee's report on Case No. 120:

"A man, thirty years old, was killed, leaving a wife and two children, four and two years of age. They had been in this country only three months. The woman spoke no English and had no trade and had no near relatives in this country, except a sister who had come over with her and was almost as helpless.

"After the fire she went to a poor cousin whose family was seriously incommoded by the addition of four people. She wished to return to Russia where she had a brother and a sister. Her passage was engaged, passport and other official papers were secured, arrangements were made for having her looked after at all points in Russia where she would change cars, and for paying her a lump sum of money with which to establish a business.

"Three days before she was to sail, however, she received a letter from her brother telling her on no account to return as there were rumors of pogroms and war. This so frightened her that she was not willing to go.

"A few days later she again changed her mind and wished to go home. Arrangements were again made for her return and again a day or two before the date of sailing she refused to go. The United Hebrew Charities was then requested to take charge of the family, $1,050 altogether has been placed with that society to be used in current expenses and in carrying out some plan for making the woman self-supporting; and $4,000 to be kept as a trust fund for the two little children."

The compassion of the gentle people on the committee breaks through the official language. They had dealt lovingly and patiently with a lost, lonely heart.

In the depth of their sorrow, some harbored pride and a different kind of need. A Red Cross investigator told of the girl in the New York Hospital who, when asked whether she needed anything, replied:

"Bring me a copy of Shakespeare and another of Tolstoy, in Yiddish."

12. PROTEST

The clangor of a sound fraught with terror.
—CANTO IX:65

IN THE city churches and synagogues, the day after the fire resounded with prayers for the dead and admonitions for the living. On this first, confusing day, the conscience of the city stirred and began to find a questioning voice in the pulpits.

At the Calvary Baptist Church, the Reverend Dr. R. S. MacArthur intoned:

"Merciful God, teach employers of labor the duties which they owe to those under their care in the proper construction of factories, in making proper exits and in all other ways caring for the comforts and especially the lives of those in their employ. If the law has been violated, may punishment swift and sure and just be speedily inflicted."

At Grace Church, on Broadway and Eleventh Street, only a few blocks from the scene of the fire, the rector, Rev. Dr. Charles L. Slattery, probed deeper. He told his parishioners that there was no need to grieve for those who had died. "Their troubles were quickly over and God will care for them." It was the survivors who merited sympathy. "One of the hard facts that will confront these bereaved," he continued, "is that it will probably be explained that the death of their loved ones was needless. It will perhaps be discovered that someone was too eager to make money out of human energy to provide the proper safeguards and protections." He hoped that the workers had not died in vain, that the tragedy "would make New York stop to think whether it was not allowing men to go too madly and disastrously and selfishly in pursuit of money."

In the week following the fire, this kind of soul-searching was repeated many times. Public-spirited citizens, community organizations, and institutions met to seek the causes of the tragedy and to determine whether each, even if only in a minor fashion, shared responsibility for it. Special meetings were held by:

The Public Safety Committee of the Federation of Women's Clubs
The Chamber of Commerce of New York
The executive committee of the Architectural League
The committee on city affairs, insurance and fire regulations of the New York Board of Trade
The board of directors of the United Cloak, Suit and Skirt Manufacturers of New York
The Merchants' Association
The Association for Improving the Condition of the Poor
The employers' welfare section of the National Civic Federation

The sense of public guilt made itself felt at all the meetings. At the National Civic Federation session, Assemblyman C. W. Phillips nailed it down when he told the businessmen that the great industrial state of New York, with its many thousands of factories, "has 75 game protectors in its Department of Game but only 50 human protectors in its Department of Labor."

The first protest meeting was held the day after the fire at 43 East Twenty-second Street, headquarters of the Women's Trade Union League. Representatives of twenty labor and civic organizations met there at the League's invitation.

League President Mary Dreier presided. Rabbi Stephen S.

Wise immediately called for the formation of a citizens' committee of twenty-five to begin the gathering of facts about the tragedy and thus provide a basis for the drafting of proposed remedial legislation. "We don't want an outburst of charity for those who have suffered only to have the whole thing forgotten in short order," the rabbi said.

Some declared that the usual kind of official investigation would not do. Others urged proceeding with caution. Leonora O'Reilly of Local 25, ILGWU, and an officer of the WTUL, replied that it was easy enough to talk of orderly, judicial procedures. "But there is little comfort in that for the thousands of women and girls who have to start work tomorrow morning in factories that are no safer—and most of them worse—than the Triangle shop.

"We cannot be calm in the face of such a frightful loss of life when it is clearly the result of the negligence of our lawmakers in the first place for making inadequate laws, and in the second place the officials for their failure to force compliance even with these," she added.

During the session, a committee composed of Mary Beard, Ida Rauh, Mrs. Stephen Wise, and Mrs. Ollesheimer drew up a questionnaire to be handed out to workers who would be told that all replies were to be held in confidence. The questions asked were:

> In your shop or factory, are the doors locked? Are there any bars on the windows? Are the freight elevator doors closed during the day? Are there fire escapes on all floors and is there free access to the fire escapes? Are there any scraps left near the motor or engine?

On Wednesday evening, thousands of relatives and friends of the victims crowded into Grand Central Palace for a memorial meeting called by ILGWU Local 25. Most of them were women.

The tenseness that gripped the audience from the start reached the breaking point during the speech by Abraham Cahan, the editor of the *Jewish Daily Forward* who had made that paper the guide and educator of the immigrant Jewish masses. Cahan began by praising Mayor Gaynor for heading the contributions to the relief fund.

But many of the workers in his audience did not share Cahan's faith in the power of a free society to persuade or curb the arrogance of wealth and they were disillusioned with men of good will who somehow were unable to affect the basic issues of poverty. At the mention of the Mayor's name, hisses broke out in all parts of the hall. He was part of the system that protected vested interests of wealth and tolerated the conditions that had caused the death of their comrades. Cahan strove to make himself heard.

He told of an excited, wild-eyed worker who had come marching into his office full of anger and protest over the tragic fire. Deprecatingly, Cahan declared: "He told me that only the placing of a few bombs in the camp of the capitalists would bring redress to the working classes."

The immediate effect was far from what Cahan had wanted, for now, according to the *Times,* the hall filled with shouts: "Throw a bomb under City Hall!" "Blow the place up!"

The meeting threatened to get out of hand. Chairman Jacob Panken rapped forcefully for order. Finally he managed, with excellent judgment, to ask for a moment of silent prayer for the dead. A low, confused murmur filled the hall.

Then a girl's voice broke in a deep sob. A woman further back in the hall sobbed, too, and sounds of convulsive weeping spread throughout the hall. A woman screamed.

Instantly, the tension snapped. Cry after cry rang out. Men and women, wailing, sprang to their feet.

The chairman could be seen speaking but he could not be heard. Many survivors were seated on the stage, and they, too, began loudly crying, calling out the names of departed co-workers. The hall resounded with a single, mass shriek of despair.

Women fell fainting on the floor.

Captain O'Connor of the East Twenty-second Street Police Station had expected trouble, and for that reason he had stationed sixty of his men in the hall. He led them down the aisles and through the rows of seats, trying to calm the throng and carrying those who had fainted out of the area where they might be trampled if the panic increased.

A police alarm had brought doctors and ambulances from Flower Hospital, and these set up emergency equipment in the lobby. More than fifty women were treated, but only one had to be taken directly to the hospital.

When quiet was again restored, Leonora O'Reilly was introduced as the next speaker. As she spoke of the "martyrs who died that we might live," the hysteria threatened to erupt once more.

A girl who had survived the fire but had succumbed to the panic at the meeting tried to force her way back into the hall, screaming and tearing her hair. The *Times* reported that several girls in the back rows tried to restrain her. Policemen ran toward her. "In the struggle, the girl's shirtwaist was torn from her back and she was carried screaming out of earshot," the paper said.

Now the audience had purged itself of its own hysteria. It listened in silence and in sorrow to the speakers. One of the last was A. M. Simons, editor of a Chicago Socialist publication. He declared that only labor could take care of its own safety.

"We have the votes. Why should we not have the power?" Simons asked. "Your future lies in unionism. Your union should have the right to decide questions which are of most concern to working people and it can get this right only by organizing. These deaths resulted because capital begrudged the price of another fire escape."

It was a thoughtful audience, rid of its earlier excitement, that filed out of the Grand Central Palace when the meeting ended. As they left, they were handed leaflets announcing a debate on the following Sunday at the Rand School on the subject: How can the Socialist victory in Milwaukee be duplicated in New York?

There was anger and determination, not hysteria, at the rally sponsored on Friday evening by the Collegiate Equal

Suffrage League and held at Cooper Union, where in February, 1860, Lincoln had campaigned against another kind of slavery. A flaring banner, stretched across the historic platform, proclaimed: "Votes for women. Locked doors, overcrowding, inadequate fire escapes. The women could not, the voters did not, alter these conditions. We demand for all women the right to protect themselves."

The three main speakers of the evening were Meyer London, who had been counsel for the striking shirtwaist makers the year before; Morris Hillquit, the famous Socialist lawyer; and Dr. Anna Shaw, the noted suffragist.

"As I read the terrible story of the fire," she told the audience, "I asked, 'am I my sister's keeper?' For the Lord said to me, 'where is thy sister?' And I bowed my head and said, 'I am responsible.' Yes, every man and woman in this city is responsible. Don't try to lay it on someone else. Don't try to lay it on some official.

"We are responsible!

"You men—forget not that you are responsible! As voters it was your business and you should have been about your business. If you are incompetent, then in the name of Heaven, stand aside and let us try!

"There was a time when a woman worked in the home with her weaving, her sewing, her candlemaking. All that has been changed. Now she can no longer regulate her own conditions, her own hours of labor.

"She has been driven into the market with no voice in the laws and powerless to defend herself. The most cowardly thing that men ever did was when they tied woman's hands and left her to be food for the flames."

Then Dr. Shaw turned to a consideration of the recent ruling by the Court of Appeals declaring unconstitutional a proposed employer's liability law.

"Something's got to be done to the law," she warned. "And if it is not constitutional to protect the lives of workers then we've got to smash the constitution! It's our 'instrument,' and if it doesn't work, we've got to get a new one!"

The audience cheered.

Hillquit, declaring that sympathy was useless although he could understand it, argued for social change. "Punishment as revenge is also natural—but useless," he declared. "I do not believe in jail as a remedy for social evils.

"The girls who went on strike last year were trying to readjust the conditions under which they were obliged to work. I wonder if there is not some connection between the fire and that strike. I wonder if the magistrates who sent to jail the girls who did picket duty in front of the Triangle shop realized last Sunday that some of the responsibility may be theirs. Had the strike been successful, these girls might have been alive today and the citizenry of New York would have less of a burden upon its conscience."

Hillquit then told the audience that more than 50,000 workers were losing their lives every year in industrial accidents—more than 1,000 a week. Then he turned on the owners of Triangle.

"Mr. Harris and Mr. Blanck were there at the time the fire broke out. They escaped. We congratulate them. My friends," Hillquit concluded, "what a tremendous difference between the captains of ships and the captains of industry!"

Meyer London recounted the story of the 1909 shirtwaist makers' strike with special attention to the part played in it by the girls who worked at Triangle. "Now," he said, "we will get an investigation that will result in a law being referred to a committee that will report in 1913 and by 1915 a law will be passed and after that our grafting officials will not enforce it."

A fourth speaker—Fire Chief Croker—had been scheduled to address the meeting. Instead, he pleaded that the press of official business made it impossible for him to do so and that he would be pleased to send a statement, which was read to the audience. It concluded as follows:

"It would be my advice to the girls employed in lofts and factories to refuse to work when they find the doors locked.

"It all comes right down to dollars and cents against life. That is at the bottom of the entire thing. Mr. Owner will come and say to the Fire Department: 'If you compel us to do this or that we will have to close the factory; we cannot afford to do it. I only get so much interest.'

"And if we say to the builders: 'You will have to put in a fire tower,' Mr. Builder will answer, 'It is impossible.' He is going to invest $15,000 in this building and he hopes to get so much a square foot and if so many square feet are cut off, he cannot build.

"It comes right down to dollars and cents against human lives no matter how you look at it."

Two memorial meetings were held on Sunday April 2, a week and a day after the fire. Early in the afternoon, the cloakmakers gathered at Grand Central Palace and under the chairmanship of Benjamin Schlesinger, who served as president of the ILGWU, marked their sorrow and their anger. But it was the second meeting, starting at three in the afternoon, that the city watched most closely. It was not a meeting of garment workers. It was not an indignation meeting of a political party. It was the first gathering together in public assembly of persons in all strata of the community who felt that a sense of outrage was meaningless unless turned into a force for reform.

Anne Morgan had rented the Metropolitan Opera House on behalf of the Women's Trade Union League. When the meeting was called to order, the stage was filled with distinguished leaders of society, government, church, and charity, as well as some who spoke for labor.

The galleries filled first, with East Siders who came early and packed the upper part of the house.

Below them were those who arrived in Sunday finery, the men in high hats and plush-trimmed overcoats, the women trailing furs and feathers. Could the gap separating the boxes from the balconies be bridged? Could such a heterogeneous audience find a single common voice? This was the important issue. Never before had a similar attempt been made to find a formula for civic reform. Even at this meeting there was no agreement—only a beginning.

Governor Dix had been scheduled to preside. In a letter to Miss Morgan he regretted his inability to attend; he was attending the funeral of Samuel J. Abbott, an aged watchman who had lost his life in the fire which that week had destroyed a part of the State Capitol office. Important records and the sword presented by Frederick the Great to George Washington had been lost.

In opening the meeting, Jacob Schiff called upon former District Attorney Eugene A. Philbin to preside. Then he reported that the relief committee had already received donations totaling $75,000. He called it "the public's conscience money."

The balconies cheered. The speakers on the stage sensed the emotion in their applause, the dangerous impatience in their booing and hissing. Each in turn tried to define a position of extreme reform without crossing over to the area of revolutionary destructiveness. Before the end of the meeting, the call was to sound for forceful change.

The Charities Director of the Roman Catholic Diocese of Brooklyn, Monsignor White, warned that "we have allowed a contradiction to grow up between our economic and our spiritual ideals; we have put property rights above life.

"The workers have a right to life and it comes before our right to the ease and luxury that flow to the community through the production of the wage earners. But industrial salvation must come from the working class itself, through its labor unions."

Bishop David H. Greer broadened the concept of sin to include failure to meet community responsibilities. "This calamity," he declared, "causes racial differences to be forgotten for at least a little while and the whole community rises to one common brotherhood. One thing is sure. Hereafter the laws as to fire protection must be enforced not for a few weeks or a few months but for all time, faithfully, continuously, and effectively. If this is not done, the responsibility —the sin—is on the public, on us."

At these ringing words, loud applause broke out. Rabbi Stephen Wise rushed forward on the stage holding up both arms in a signal that silenced the audience.

 "Not that! not that!" he shouted. "This is not a day for applause but for contriteness and redeeming penitence.

"It is not the action of God but the inaction of man that is responsible. The disaster was not the deed of God but the greed of man. This was no inevitable disaster which could not be foreseen. Some of us foresaw it.

"We have laws," Rabbi Wise continued, "that in a crisis we find are no laws and we have enforcement that when the hour of trial comes we find is no enforcement. Let us lift up the industrial standards until they will bear inspection. And when we go before the legislatures let us not allow them to put us off forever with the old answer, 'We have no money.' If we have no money for the necessary enforcement of the laws which safeguard the lives of workers it is because so much of our money is wasted and squandered and stolen."

The applause which the rabbi had deprecated roared forth from orchestra and gallery alike. When he could be heard again, he told the audience that the purpose of the meeting was not to placate with charity but to find redress in justice, remedy in prevention.

The rising political temper of the meeting could be measured by the cheering that followed the statement of Professor E. R. A. Seligman, the noted economist from Columbia University: "We are weltering in a chaos in this city with a display of anarchy—of administrative impotence."

But in the end those in charge of the meeting asked for the adoption of a resolution. It was a good resolution, calling for the creation of a Bureau of Fire Prevention, asking for more inspectors, demanding the creation of a system of workmen's compensation.

But it was only a resolution. Many who heard it were tired of resolutions and had no faith in reform. Hisses sounded in the galleries. The loudest cheers were not for the reform program but for the dissenting speakers from the audience who said that citizens' committees had never been able to accomplish anything, that there would be no improvement for the working classes until in class solidarity they demanded it at the polls and through committees of their own. Some called for political organization of the workers. Others asked that labor union officials be appointed as fire inspectors.

The shouts from the galleries grew louder as did the hissing of those who tried to counter from the orchestra and boxes. The meeting threatened to break up in disorder.

It would have, if not for a speaker who suddenly gripped the audience with her words and held it spellbound. She was slight Rose Schneiderman, her long braided hair showing in a huge bun under her hat. She had been a leader in the strike at Triangle. She had seen her girls beaten and jailed; now she had seen them burned and dead. She spoke hardly above a whisper.

"I would be a traitor to those poor burned bodies," she said, choking back her tears, "if I were to come here to talk good fellowship. We have tried you good people of the public —and we have found you wanting.

"The old Inquisition had its rack and its thumbscrews and its instruments of torture with iron teeth. We know what these things are today: the iron teeth are our necessities, the thumbscrews are the high-powered and swift machinery close to which we must work, and the rack is here in the firetrap structures that will destroy us the minute they catch fire.

"This is not the first time girls have been burned alive in this city. Every week I must learn of the untimely death of one of my sister workers. Every year thousands of us are maimed. The life of men and women is so cheap and property is so sacred! There are so many of us for one job, it matters little if 140-odd are burned to death.

"We have tried you, citizens! We are trying you now and you have a couple of dollars for the sorrowing mothers and brothers and sisters by way of a charity gift. But every time the workers come out in the only way they know to protest against conditions which are unbearable, the strong hand of the law is allowed to press down heavily upon us.

"Public officials have only words of warning for us—warning that we must be intensely orderly and must be intensely peaceable, and they have the workhouse just back of all their warnings. The strong hand of the law beats us back when we rise—back into the conditions that make life unbearable.

"I can't talk fellowship to you who are gathered here. Too much blood has been spilled. I know from experience it is up to the working people to save themselves. And the only way is through a strong working-class movement."

There had been no interruptions. Now there was no applause. She finished and turned and walked back in dignity and in sorrow to her seat in the front row on the stage.

The challenge was being formulated.

The meeting finally passed its resolution, appointed its committee, expressed its sorrow, resolved to act. But on the East Side where the loss was felt in blood and flesh, the sorrow stayed deep and silent.

One among them spoke for them. The Jewish workers called Morris Rosenfeld the poet laureate of the slum and the sweatshop. He put his tears and his anger into a dirge, and four days after the fire, the *Jewish Daily Forward* printed down the full length of its front page:

Neither battle nor fiendish pogrom
Fills this great city with sorrow;
Nor does the earth shudder or lightning rend the heavens,
No clouds darken, no cannon's roar shatters the air.
Only hell's fire engulfs these slave stalls
And Mammon devours our sons and daughters.
Wrapt in scarlet flames, they drop to death from his maw
And death receives them all.

Sisters mine, oh my sisters; brethren
Hear my sorrow:
See where the dead are hidden in dark corners,
Where life is choked from those who labor.
Oh, woe is me, and woe is to the world
On this Sabbath
When an avalanche of red blood and fire
Pours forth from the god of gold on high
As now my tears stream forth unceasingly.
Damned be the rich!

The Triangle Fire

Damned be the system!
Damned be the world!

Over whom shall we weep first?
Over the burned ones?
Over those beyond recognition?
Over those who have been crippled?
Or driven senseless?
Or smashed?
I weep for them all.

Now let us light the holy candles
And mark the sorrow
Of Jewish masses in darkness and poverty.
This is our funeral,
These our graves,
Our children,
The beautiful, beautiful flowers destroyed,
Our lovely ones burned,
Their ashes buried under a mountain of caskets.

There will come a time
When your time will end, you golden princes.
Meanwhile,
Let this haunt your consciences:
Let the burning building, our daughters in flame
Be the nightmare that destroys your sleep,
The poison that embitters your lives,
The horror that kills your joy.
And in the midst of celebrations for your children,
May you be struck blind with fear over the
Memory of this red avalanche
Until time erases you.

13. DIRGE

. . . all together they withdrew,
Bitterly weeping.

—CANTO III:106

THE WEEK following the fire was filled with funerals. Long, wailing processions moved through the labyrinth of crowded, narrow East Side streets. "Every block in the section south of Tenth Street between Second Avenue and Avenue C seems to have lost some one," said the *World*.

The crowds grew almost accustomed to the passing hearses. The bargaining at the pushcarts along the curbs stopped as they passed. In the silence, some men bared their heads.

One East Side undertaker established a record by conducting eight funerals simultaneously. On Wednesday, three funeral processions crossed. In the ensuing confusion some, for a time, followed the wrong hearse. The next day, an old Jewish sexton sat beside the driver of a hearse and dispelled confusion among the mourners by holding aloft a large cardboard sign hand-lettered: "This is the funeral of Yetta Goldstein."

There were grand funerals and there were small, sorrowful departures.

Jennie Franco had been only fifteen years old. She was laid out in the small front room of her home at 342 East Eleventh Street, next to the room in which she had been born. One hundred men and women, girls and boys, together with a brass band preceded the hearse which was followed by sixty carriages. The girls were Jennie's schoolmates. The twenty-five floral pieces were sent by friends and a carriageload of flowers came from the three societies of which her father was a member—St. Angelo di Brolo, Sons of Italy, and the St. Stephen Cammastra Order.

But those sent to their final resting places from union headquarters on Clinton Street went without bands and without flowers. From the union hall, Julia Rosen set out for her last journey followed by her three orphaned children and a group of weeping neighbors. Out of this place they carried the burned body of Essie Bernstein, and as the casket was placed in the hearse, a bearded Jew mounted to the top of a nearby stoop and intoned in the sad, angry voice and language of the Hebrew prophets: "Our poor children go to work in fire traps to avoid a life of shame. When they come home, they go to sleep in tenements which are also fire traps. By day and by night—they are condemned!"

Not all the dead went alone.

Rosalie Maltese and her sister, Lucia, went together, Bettina Miale and her sister, Frances, went together. So did Sara and Sarifine Saricino, sisters.

Sophie Salemi and Della Costello lived in neighboring houses on Cherry Street. They had worked at adjacent machines in Triangle. A score of carriages followed the two white hearses. The fifty girls who marched at the head of the procession were members of the Children of Mary Society. It was altogether fitting that they shared a funeral. These two had leaped from the ninth floor, their arms around each other.

When there were no more funerals and only the unidentified bodies in the morgue remained, the Women's Trade Union League and Local 25, ILGWU, through its manager, Abraham Baroff, applied to Commissioner Drummond for permission to hold a funeral for the nameless ones—the unclaimed dead.

The Hebrew Free Burial Society opposed a public funeral; the sense of bereavement went so deep that such a move seemed out of place. Commissioner Drummond opposed it because he feared the hysteria it would arouse.

Nevertheless, plans for a public funeral went forward. At a meeting on March 29, the committee of the League and the garment workers' union decided that only union banners draped in black were to be allowed. There were to be no other banners, no bands, no propaganda signs. The next day Morris Hillquit asked Coroner Holtzhauser to release the bodies. But the Coroner, hoping for additional identifications, insisted on holding the bodies for five days more.

On the same day, Mayor Gaynor announced he had been informed by Commissioner Drummond that the city owned a plot in the Evergreen Cemetery in the East New York section of Brooklyn. The city, he said, would inter the seven unidentified bodies in that cemetery on April 5. It would not release them for a public funeral.

The committee immediately issued a call for a funeral parade of the city's workers on Wednesday, April 5, starting at one o'clock.

On the preceding Friday and Saturday, several hundred teen-aged youngsters, including about seventy-five members of the Young People's Socialist League, distributed thousands of leaflets in the factory sections of the city. The tri-lingual handbills carried an appeal in English, Yiddish, and Italian for all workers to "join in rendering a last sad tribute of sympathy and affection." For those unable to make the trip to Manhattan there was to be a separate parade in the Brownsville section of Brooklyn.

On the day of the funeral—April 5—"the skies wept," said the *World*. All through the day, "rain, ever and again, descended in a drenching downpour."

The committee had arranged for the parade to assemble and start simultaneously at two separate locations at 1:30 P.M. The two sections, one moving out of Seward Park at the point where East Broadway and Canal Street meet in the heart of the East Side and the other heading south on Fourth Avenue from Twenty-second Street, were to meet in Washington Square and become a single line of march.

As early as nine o'clock in the morning, small groups began to assemble in Seward Park. By noon, the crowd had grown large enough to block traffic. There was a great deal of confusion until a little before one o'clock, when a single empty hearse, bedecked with flowers and drawn by six white horses covered with black nets, headed into the crowd. With a murmuring sound, the waiting people closed in around it.

Along each side of the double line of horses were eight girls in mourning dress. About four hundred relatives of victims and survivors took their positions behind the hearse. By 1:30 P.M. the downtown contingent of the funeral was ready to march.

Uptown the crowd had begun to assemble at noon from Nineteenth to Twenty-second Streets on both sides of Fourth Avenue. This division consisted chiefly of girls from the shops and factories and women from the Suffragist and Trade Union League groups. The marshals were T. J. Curtis of the Central Federated Union, J. W. Roberts of the Bonnaz Embroidery Workers' Union, Julius Gerber of the Socialist Party, and Arthur Carote, Italian organizer for the WTUL.

When the bells in the Metropolitan Life Insurance tower sounded 1:30 the first group marched out of Twenty-second Street onto Fourth Avenue. It headed south, first to Union Square, then down University Place to the northeast corner of Washington Square.

"The division was composed largely of women, most of whom, for some unexplained reason, did not wear hats or overshoes," the *Tribune* reported. "They had been standing in the streets for an hour. Women held babies and watched from the curbstone. Most men and women in the procession wore a white badge on which was printed in black letters, 'We mourn our loss.' Inspector Sweeney was in charge of the police arrangements and the best order prevailed."

The downtown contingent had also begun its march at 1:30. All four marshals of this contingent were from the Waistmakers' Union, Local 25, ILGWU. They were B. Witashkin, S. Liebowitz, A. Miller, and Louis Schaeffer. They signaled the driver of the empty hearse to start his horses.

A massed line formed behind the flower-laden wagon; the crowd grew silent. A special detail of policemen had been assembled under the command of Inspector Schmittberger to keep order. "There was something ominous about the gathering," said the *American.* "It was so silent and it was to march through a section of the East Side—the thickly populated foreign districts—where emotions are poignant and demonstrative. The police were plainly worried."

Headed by the empty hearse, the procession moved silently, slowly on East Broadway. It turned north into Clinton Street then west again into Broome Street. In the canyon of tenements, windows were crowded with women, children, and old people. As the flower-loaded, empty coach came into view, many in the windows slowly waved handkerchiefs. As the hearse passed, the cries burst forth, moving along like a rolling wave in the same slow progress as the hearse.

Where it had passed, there was silence again and windows suddenly empty. Thousands came out of the dark hallways and into the rain and the line of march. The procession lengthened as it moved.

At two o'clock, the empty hearse reached the corner of Broome and Mott Streets. Here the way was blocked by fire engines and a tangle of fire hoses. There was a fire in a tenement. All along the line, back into Clinton Street, the marchers halted. The rain fell more heavily.

The uptown division moved south at a faster pace.

The *Tribune* reported that the steady downpour did not divert the girls in the uptown contingent who were without umbrellas, hats, or overshoes from "their determination to show public honor to fellow workers who had perished." They marched "with an apparent grim satisfaction that in a sense they could wear sackcloth and ashes in so good a cause."

If the day had been filled with sunshine, the *Tribune* continued, "there would have been the chance to minimize the intensity of feeling existing among the marching members of the more than 60 sympathizing unions. Only a high devotion and sense of duty could be responsible for this protest.

"Low hanging clouds and fog shrouded the tops of buildings. The Metropolitan tower was invisible above its clock. There was the suggestion of smoke in the atmosphere. The streets were filled with puddles of water. Women in lamb's wool coats, accustomed to ride in automobiles, were splashed by passing vehicles as they trudged along in the beating rain. Hundreds of thousands stood on the sidewalks, their umbrellas appropriately indicating an unbroken border of black. Policemen, mounted and afoot, wore regulation black raincoats."

The *World* noted that "the street mud oozed through their thin shoes; they marched on silent, uncomplaining." The *Tribune* added that "here and there an older woman would insist upon taking a thinly clad young girl under her umbrella, or, if clad in a wool coat, under her wing."

The two contingents moved toward each other for the joining at Washington Square. The uptown division reached the park at two o'clock. For the next hour, it waited in the rain for the arrival of the downtown division. Shortly after three o'clock, the empty hearse heading that section came into view as it moved up Macdougal Street to the southwest corner of the Square. The two divisions of the city's massed workers were now facing diagonally across the park.

Washington Place, from the Square to Mercer Street, was one solid gathering of people. When word spread that the two sections of the parade had arrived, the crowd somehow managed to open a narrow lane down the middle of the street, expecting the parade to file past the Asch building.

But Inspectors Schmittberger and Sweeney had misgivings. They feared an outburst of mass hysteria and after a short conference altered the route of the procession so that the downtown divisions, instead of joining the other for a march through Washington Place would meet it at Fifth Avenue and proceed uptown.

The change had been made because, as the *American* reported, "it was not until the marchers reached Washington Square and came in sight of the Asch building that the women gave vent to their sorrow.

"It was one long-drawn-out, heart-piercing cry, the mingling of thousands of voices, a sort of human thunder in the elemental storm—a cry that was perhaps the most impressive expression of human grief ever heard in this city."

At almost the same moment the cry filled the air at Washington Square park a third procession started from the morgue. For an hour, eight black hearses, each drawn by a span of black horses and hung with white drapery, had been stationed in front of the morgue gate.

Inside, attendants had spent the morning preparing the unidentified bodies in white linen shrouds. Each was then placed in a casket of black broadcloth with silver handles. On the cover of each casket was a silver plate upon which was engraved: "This casket contains a victim of the Asch building fire. March 25, 1911."

A silent crowd, standing in the steady drizzle, filled the sidewalks along First Avenue from East Twenty-sixth Street to East Twenty-third Street. A few minutes after three, they saw the eight hearses, led by a cordon of mounted police, go by.

Behind the policemen was an open carriage spilling over with flowers. Each casket, in its hearse, was covered by a wreath of roses entwined with orchids. The eight hearses were in two ranks of four.

Commissioner Drummond, his son Walter, First Deputy Commissioner Goodwin, Dr. Walter Conley, and William Flanagan, architect for the Charities Department, constituting the official party, brought up the rear in two automobiles. There were no carriages and no mourners following.

At the Twenty-third Street Ferry, the procession boarded the ferryboat *Joseph J. O'Donohue,* headed for Greenpoint in Brooklyn. Just as the gates were being lowered, more than a hundred mourners who had been in the waiting room more than an hour, rushed aboard. One in this group was Mrs. Carrie Lefkowitz. She was dressed in black.

She said: "I have lost my sister, Minnie Mayer. Every day, since the fire, I visited the morgue. Every day, I peered into those poor burned faces but my sister I cannot tell. I feel that one of those in the wagons is my sister. So I follow them to the cemetery."

When they debarked from the ferryboat, some of the mourners boarded streetcars. Others stood studying the rain-filled sky. One reporter spotted a white-bearded patriarch leaning on the arm of a young girl. He heard her ask: "Shall we ride, father?"

He lifted his arms and his face as if to feel the rain on them and replied in a loud voice to the girl: "It is written that we shall follow our dead on foot, even unto the grave."

Following the empty hearse, the first marchers from the downtown contingent moved under the Washington Arch and headed up Fifth Avenue at 3:20 P.M. They marched eight abreast.

"From that minute, until 6 o'clock, a steady stream of marchers passed under the arch and every moment of the time the rain fell with unrelenting steadiness," said the *Herald*. "A few figures watched from the upper stories of the old brick mansions in Washington Square North and lower Fifth Avenue."

As promised, there were neither bands nor banners in the line of march. Only one single streamer appeared, and it was carried in the division of the parade made up of women's garment workers. It read: "We Demand Fire Protection."

"Now and again," said the *Herald,* "elderly women in the line gave vent to their feelings in a wailing chant to which marchers and spectators listened with awe and reverence. But the greater part of those in line moved along saturnine and silent. None fell out of line."

Ambulances with attending doctors and internes had been stationed along the route of the march. No patients were treated. "I fear that pneumonia will claim more victims than the fire," said one attendant stationed near the arch.

In the Evergreen Cemetery, they buried the nameless ones.

Hundreds had waited in the heavy rain at the entrance to the cemetery. With silence, bared heads, tears, they honored the dead as they passed. But once again, as at the morgue, there were the curious and the disorderly. As the hearses moved into the cemetery, "half-grown boys ran at top speed through the grounds, jumping over and on graves," the *Tribune* reported.

A pit 15 feet long had been dug. The eight coffins were placed alongside of it. At the end of the pit a small tent had been erected.

The small group of city officials huddled in the rain in front of the tent. Commissioner Drummond expressed the sorrow of the city. Then Monsignor William J. White read the Catholic service over one body. Father William B. Farrell made the responses.

The Episcopal burial service was read by the Reverend Dr. William B. Morrison over another body.

After that, Rabbi Judah L. Magnus spoke ancient Hebrew words.

Architect Flanagan had drawn up a plan for the burial plot that would make future identifications possible. He stood over the pit and pointed out each place as the caskets were lowered:

No. 46
No. 50
No. 61
No. 95
No. 103
No. 115
No. 127

The eighth casket had neither name nor number. "It contained the dismembered fragments picked up at the fire by the police and unclaimed," said the *Herald*.

When the last casket had been lowered, Peter J. Collins, Frank Corbett, John Lloyd Wilson, and James J. Byrne stepped out of the crowd. They comprised the quartet from Elks Brooklyn Lodge No. 22. They ended the service by singing "Abide with Me" and "Nearer, My God, to Thee."

The marchers continued up Fifth Avenue.

Rose Schneiderman, the slight, red-haired girl who had stirred the meeting in the Metropolitan Opera House, now stirred those watching the march of sorrow:

"Little Miss Schneiderman, hatless and without a raincoat, tried to trudge along in the dripping procession," the *Times* wrote. "But long before it reached its uptown destination, she began to falter." Help came from Mary Dreier, WTUL president, and Helen Marot, the League's secretary. Seeing her plight, "each took her under an arm and the three leaned into the wind supporting each other."

The leaders of the parade claimed it was "the largest demonstration ever made here by working people." First official anticipations of the size of the demonstration, considering the rain, were low.

While the parade was in progress, the estimates were lifted from 100,000 to 120,000. The final police estimate was that about 400,000 had seen the parade and of these about one-third had marched in it.

At Thirty-third Street, the marchers turned east to Madison Avenue, then south to Madison Square Park, where they disbanded at the base of the Metropolitan tower. At the Square, a reporter asked Rose Schneiderman if she felt sick.

"Oh please do not speak of my feeling ill," she replied. "We don't know how sick the other girls were. This parade has been the only thing that will demonstrate to the people the enormous responsibility resting on them to see to it that fire protection is given these thousands and thousands of factory workers.

"As we marched up Fifth Avenue, there they were. Girls right at the top of hundreds of buildings, looking down on us. The structures were no different from the Asch building in a majority of cases with respect to the lack of protection against fire; many were in a far worse condition. There they were, leaning out of the upper windows, watching us. This, not the rain, is making me sick."

 Rose Schneiderman remembers: "As we disbanded, we

heard the bells of the Metropolitan Life tower. It was dark and raining so we couldn't see the top of the tower. But we could hear the bells. That night we were certain they tolled for us."

14. SHIRTWAIST

With the fist closed . . .
—CANTO VII:57

ISAAC HARRIS, and Max Blanck, known to the trade as "the shirtwaist kings," were indicted on charges of manslaughter, first and second degree, in mid-April.

They had moved their firm into the Asch building in 1902, occupying at first only the ninth floor. In one of the nation's most competitive businesses, they had steadily risen toward the top. In 1906, they took over the eighth floor of the Asch building. In 1908, the volume of their production in shops in New York and in Philadelphia hit the million-dollar mark, and soon they established their general office on the tenth floor of the Asch building.

Both partners had imagination as well as ambition, and each was expert in the portion of company responsibilities undertaken by him. Harris, the inside man, knew all about garment production, machinery, how to keep the work flow going through the plant. He was of medium height, with a serious expression, and his daily task was to patrol the factory, moving impressively down its aisles, checking, questioning, directing.

Blanck, on the other hand, was the sporty type, and as he was the "outside man," this was of considerable help in developing strong ties with the buyers for stores, who were always well entertained by Triangle. He, as well as Harris, owned a large automobile and used a chauffeur. Only a few days before the fire, Blanck had won a case against Sol Lichtenstein, identified as a bookmaker, to whom he had paid a check for $875 "for losses in the racing game," the *Tribune* reported. Lichtenstein had endorsed the check and passed it on to someone else. In the meantime, Blanck had stopped payment on the check, leaving the bookmaker out on a limb. In City Court, the judge agreed with Blanck, who took the stand that the check was for a gambling debt and was therefore not collectible by court order.

Harris and Blanck showed the same kind of shrewdness during the strike that hit their factory late in 1909. For the faithful who stayed in the shop they installed a phonograph on the ninth floor. One who was there remembers that while the pickets on the street "were being beaten, we danced during lunch time. Mr. Blanck even used to give out prizes to those who were the best dancers. But once the strike was over—no more dancing, no more prizes, no more phonograph."

Now, with the threat of punishment, perhaps even years of imprisonment, hanging over their heads, the Triangle partners sought to mend their tarnished public image. In the weeks after the fire they undertook their first advertising campaign in New York newspapers.

Between March 31 and July 13, they sent twenty-five contracts for advertising, for a total expenditure of $4,801.50 to the papers, including some foreign-language publications. A contract, with a check, was offered to the New York *Call,* which photographed the check, returned it, and then ran an account of the transaction in its columns. The *Sun* and the *Catholic News* also returned the checks; the *Morning Telegraph* said it had never run the advertising covered by the check.

Harris and Blanck had made their fortunes manufacturing shirtwaists for ladies. They were the largest firm in the business, and the garment they made was aimed at mass sales. It was of medium quality and sold for $16.50 to $18 a dozen —wholesale.

More than any other item of feminine apparel, the shirtwaist symbolized the American female's new-found freedoms. It was a cool and efficient bodice garment, generally worn with a tailored skirt. Early in the new century it became standard attire for thousands of young ladies taking positions with industrial and commercial enterprises. The popular artist, Charles Dana Gibson, immortalized the shirtwaisted female. He pictured her as a bright-eyed, fast-moving young lady, her long tresses knotted in a bun atop her proud head, ready to challenge the male in sport, drawing room, and, if properly equipped with paper cuff covers, even in the office.

The shirtwaist was topped at the neck by a recognizable variant of an open or buttoned mannish collar. In sharp contrast to the masculine stringency thus achieved, it descended in a broad expanse across the bosom, then by means of tucks, darts, or pleats tapered dramatically to a fitted waistline. The secret of its perennial popularity was in its lines and the fabric of which it was generally made. Crisp, clean, translucent—and more combustible than paper—the sheer cotton fabrics produced opposite but pleasing effects. The *bouffant* quality of the fabric enhanced the figure it enfolded. At the same time its sheerness piquantly revealed the dainty shape beneath.

At the start of the new century the garment industry began a move out of the slum workshops and railroad flats in which entire families labored over bundles of cut garments farmed out to them by jobbers. By 1903, Ernest Poole, later winner of the first Pulitzer Prize for fiction, was able to report that in New York, 70 per cent of coat production was done in factories in contrast to the old farming-out system. Poole pointed out that the factory meant "endless saving, dividing, narrowing of labor. The worker's strength is no longer wasted in pushing a treadle; the machine is run by power. Each worker does one minute part swiftly, with exact precision."

The advantages of locating in one of the new towering loft buildings were described by Arthur E. McFarlane as including "cheaper insurance because loft buildings were fireproof. The installation of motors and shaftings allowed manufacturers to use electricity instead of the old gasoline engines and electricity was cheaper. There was daylight until 5 o'clock, even in winter, which meant a saving of gaslight."

But the more important advantage arose from the New York factory law requirement that each worker be provided with 250 cubic feet of air. In the new buildings, the ceilings were higher than in the old with the result that more workers could be crowded into a given area. In terms of square feet per worker, the new factories provided less, not more space for each employee.

Because of this, firms like Triangle could, in effect, draw together under one roof the scores of homework units and sweatshops they would have had to utilize under the old farming-out system. Instead, these now became self-organized teams of workers.

Triangle would hire a good machine operator and allocate to him half a dozen machines out of the 240 on the ninth floor. In turn, this operator, in reality a contractor for the firm, would hire the young girls, immigrants and women from his home town across the sea, as learners. He would teach them how to make the separate parts of the garment which he, as master craftsman, would join together.

Mary Leventhal kept the record of cut work distributed and the finished waists returned by the leader. Only he knew the value of the work done by his team because only he had bargained with the company for the rate on each style.

"The girls never knew," says Joseph Flecher. "For them there was no fixed rate. They got whatever the contractor wanted to pay as a start. In two or three weeks they knew how to sew very well. Never mind. For a long time they still got the same low pay. Triangle and the inside contractor got the difference."

The company dealt only with its contractors. It felt no responsibility for the girls. Its payroll listed only the contractors. It never knew the exact total of its workers.

On Saturdays, it was the inside contractor who paid the girls. In each case he took into account the skill and speed, the family relationships or the defenselessness of the individual worker. She, in turn, showed her appreciation by being docile and uncomplaining.

But the gathering of the workers under one roof created a new circumstance which, as Poole pointed out, "is a help to the workers. In a system of small, scattered shops, the union had no chance." Gathered together, workers found it easier to organize and remain united. They could see the real employer beyond the contractor with whom they dealt. "The workmen are learning to strike it together," Poole said. So, too, were the working women.

For in time dissatisfaction grew among the girls which resulted in pressure on the contractors. But unless the contractors received more pay from the employers, they could not pass on more pay to their own teams. The Triangle partners sensed the increasing uneasiness.

Bernard Baum remembers the first serious flare-up that eventually led to the strike at Triangle. About a year and a half before that walkout, he says, Jake Kline and Morris Elzufin, "two soft-spoken group leaders, found they had little left from their pay for themselves after paying off their teams. They got a curt answer when they went to appeal to Bernstein, the manager. He ended up telling them to get back to their machines, finish their work, and get out.

"Five minutes later, one of Bernstein's deputies—Morris Goldfarb, the toughie—and a few others of his kind approached the two men at their machines, broke the threads of their cotton spools—Bernstein's orders, they said—and told the two men to go."

Kline refused. He protested he had work to finish, that he wouldn't get paid if he left it unfinished.

"When Bernstein heard Jake's hollering, he came running, grabbed him by the back of the neck and started to drag him to the door, slapping him left and right. Goldfarb grabbed Elzufin."

With his shirt torn and his glasses broken, Jake managed to twist free for a moment and shouted into the shop, where all work had stopped: "People . . . workers . . . look what they are doing to us . . . get up from your machines!"

They did. The wheels kept turning, but no work passed under them. Only scattered groups remained when the turmoil stopped. Most of the operators had gone to the street, where they gathered around Jake and Morris.

Some went directly to the headquarters of ILGWU Local 25 to ask for help. But far from being able to help, the garment union was in need of help itself. It had only 400 devoted members, and virtually no funds. Reluctantly, it advised conciliation.

After the Triangle workers returned to their machines, management sought ways in which to undercut or block the rising resentment. It hit on the idea of forming a company union.

This became an exclusive group in the factory, enrolling only one of every six workers. But instead of dampening the discontent, it fed the anger of those excluded from its ranks, who began to call in small groups at union headquarters for help.

Survey magazine reported that "discontent grew even among the members of the company society," so that when a meeting of Triangle workers was held at Clinton Hall in September, 1909, all but seven members of the company union were present. Word got back to the firm, and on Friday, September 24, the day after the meeting, "the employers called the girls together and expostulated with them more in sorrow than in anger. Terms were once more arranged between a delegation of operators and the firm and the next day everyone went back to work as usual.

"On Monday, however," the magazine report continued, "when the girls reported for work the shop was found closed. The next day it was once more open. But no union girls were taken back so, within 36 hours, through the agency of the society whose dwindling membership then numbered exactly seven—all of them sisters, cousins and aunts of the members of the firm—the lockout began."

Local 25 immediately declared a strike against Triangle. The walkout surprised the city because few had expected immigrant Jewish and Italian girls to strike as effectively as these did. Triangle was hard pressed for production. On October 29, it wrote to other blouse manufacturers urging them to join in the formation of an employers' protective association to combat the spreading evil of unionism.

The picket line held firm and defectors were only those who found employment in other shops. In its November 13 issue *Survey* magazine reported that the strike of two hundred women employees at Triangle was being conducted according to approved strike methods.

"The management of the company has tried to protect its rights against the girls by calling into commission a regiment of police, plainclothes detectives and other burly men who, the girls declare, are nothing more than neighborhood thugs, sufficient in number to handle a general industrial disturbance, willing to strike women and to hustle them off to court on flimsy pretexts.

"The girls have been entirely orderly but police interference has made them appear otherwise," the magazine report continued. "The officers break in upon any who are talking together; men loafing about in the employ of the company have insulted the girls; and the least refusal to answer the officers is made excuse for prompt arrest. Unfair treatment has not stopped there for in court the judges railroaded through a whole batch of girls at a time without as much as a hearing."

Triangle had allies—at a price. Joseph Flecher recalls that "you could get a man on the beat to look away by giving him a box of cigars with a $100 bill in it. Then the hoodlums hired by the company could do their work without interference. They couldn't hit women, even on the picket line. So they brought their lady friends—prostitutes. They knew how to start fights."

The Triangle strikers also had friends but of a very different kind. Chief among these was the group of wealthy women who, in 1904, had formed the Women's Trade Union League in order to work with their less fortunate sisters in the shops and the factories for a fair deal as workers and as women. They helped "man" the picket line at Triangle. Their intervention on behalf of the girls drew public attention to the industrial conflict. In groups, they formed voluntary patrols, marching with the strikers early in the morning and late in the evening and accompanying those arrested to court to testify to their innocence.

The peak of strike publicity was reached with the arrest of Mary Dreier, WTUL president. *Survey* magazine reported that "Miss Dreier was discharged upon arrival at the nearest station house and the police attitude toward the women was deliciously revealed when the officer in charge upbraided her for not having told him she was 'the working girls' rich friend,' had he known which, of course, he would not have arrested her."

The dissatisfaction and unrest that had led the Triangle girls out onto the picket line filled the city's other shirtwaist shops. Thousands prepared to follow their example. The general walkout was touched off at four overflow meetings held on the evening of November 22 at Cooper Union, Beethoven Hall, Astoria Hall, and the Manhattan Lyceum, all in the East Side area.

The main speaker in the Great Hall of Cooper Union was Samuel Gompers, president of the American Federation of Labor. He told the audience, sensing the approach of a dramatic decision, that he had "never declared a strike in all my life. I have done my share to prevent strikes. But there comes a time when not to strike is but to rivet the chains of slavery upon our wrists.

"Yes, Mr. Shirtwaist Manufacturer!" Gompers thundered, "it may be inconvenient for you if our boys and girls go out on strike. But there are things of more importance than your convenience and your profit. There are the lives of the boys and girls working in your business."

Before the next speaker could be introduced, a slim, pretty youngster who had been beaten on the Triangle picket line, rose and asked for the privilege of addressing the gathering. When this was granted, Clara Lemlich said there had been enough talk. She called for a vote to authorize a general strike. Her impassioned plea was heard in earnest silence by the audience. They knew the hardships of striking.

The vote was overwhelmingly in favor of a strike. The chairman called on the audience to rise. He asked all to raise the right hand and when this was done he led them in the solemn Biblical oath: "If I forsake thee, may this arm wither. . . ."

The strike spread to all the shirtwaist shops in the city. Newspapers reported, at first with astonishment and then with growing sympathy, the bravery of the immigrant girls and young men who put on their best clothing to march on the picket lines.

The determination of the strikers and their friends increased as the number of arrests rose. Not at all unusual was the dramatic appearance of Mrs. O. H. P. Belmont in night court at 2:30 in the morning of Sunday, December 14.

The court was hearing the cases of those arrested late Saturday afternoon. The doors swung open and in sailed Mrs. Belmont wrapped in furs and wearing a hat so huge that it had to be anchored to her carefully coifed hair by six jeweled hat pins. While spectators and attendants gasped, recognizing her from pictures they had seen in the newspapers, she drew up in front of the bench and confronted Magistrate Butts. She was there, she told him imperiously, to stand bail for four of the girl strikers. The amount of bail was $800. She was putting up as security the Belmont mansion at 477 Madison Avenue.

When the Magistrate facetiously asked if it was sufficient to cover the amount, Mrs. Belmont fixed him with a cold stare and replied: "I think it is. It is valued at $400,000. There is a mortgage of $100,000 on it which I raised to help the cause of the shirtwaist makers and the women's suffrage movement."

Carola Woerishoffer, young, wealthy, dark-eyed, and a graduate of Bryn Mawr, did it her own way. She used her money to buy houses. Then she haunted the entrance to the Jefferson Market Court at Sixth Avenue and Tenth Street and whenever another group of arrested strikers was marched before a magistrate, she was with them in front of the bench armed with a deed, ready to slap it down for their release.

Mrs. Belmont and the other women who had joined in starting the Colony Club patronized by the "cream of the Four Hundred," including Anne Morgan, Elizabeth Marbury, and Mrs. J. B. Harriman, held a meeting with the strikers in their exclusive quarters. The *Call* reported how the strikers, "some of them children, faced the 350 women, representing the richest people in the world, sitting on the edges of their gilt chairs in the beautiful auditorium of their sumptuous club. They told how they worked for as little as $3 a week."

Said one of the youngsters: "They hired immoral girls to attack us and they would approach us only to give the policemen an excuse to arrest us. In two weeks, eighty-nine arrests were made. I too was arrested and the policeman grabbed me by the arm and said such insulting words that I am ashamed to tell you."

By Christmas Day, 723 arrests had been made. Conscientious community leaders rallied to the defense. Professor Seligman of Columbia, Lillian D. Wald, and Mary K. Simkhovitch, pioneers of social work, and Ida M. Tarbell, who had started the muckrake school of journalism with her exposé of the Standard Oil Company, were among those who publicly charged in a joint declaration that there was "ample evidence to warrant the statement that the employers have received cooperation and aid from the police," that the strikers had been "subjected to insult by the police," and that they frequently had been "arrested when acting within their rights."

Magistrate Olmstead, sentencing a striker, had lectured her: "You are on strike against God." The Women's Trade Union League wrote to the noted British socialist and playwright George Bernard Shaw, asking for a comment. Back came the reply: "Delightful. Medieval America always in the intimate personal confidence of the Almighty."

The majority of the public sided with the girls. "From the start," Ida M. Tarbell explained, "a large body of people who had perhaps thought very little on the subject of unionism and who had little acquaintance with the conditions under which shirtwaist workers lived, immediately came to their assistance.

"They said very rightly that these girls had a right to make an orderly stand to improve their conditions. In doing that, they are working not merely for themselves but for society as a whole. Anything which improves their condition," Miss Tarbell concluded, "must improve everything in town."

The strike ended after thirteen weeks. By that time ILGWU Local 25, which had entered the walkout with four hundred members and $10 in its treasury, had signed contracts with 354 firms.

Miriam Finn Scott listed some of the gains won by the strikers as the establishment of a 52-hour work week, wage increases of 12 to 15 per cent, a promise to end the subcontracting system in the shops, a 2-hour limit on night work, and "in the slack season to divide the work among all workers instead of giving it to the favored few."

In the framework of that time, progress had been made. But the strike that had started at Triangle was not won at Triangle.

The girls who had worked at the Triangle shop "were beaten in the strike," Martha Bensley Bruere declared. "They had to go back without the recognition of the union and practically no conditions.

"On the 25th of March," she continued, "it was the same policemen who had clubbed them back into submission who kept the thousands in Washington Square from trampling upon their dead bodies, sent for ambulances to carry them away and lifted them one by one into the receiving coffins."

Rose Safran was one of the Triangle strikers who returned in defeat. "I was one of the pickets and was arrested and fined several times," she said. "The union paid my fines. Our bosses won and we went back as an open shop having nothing to do with the union. But we strikers who were taken back stayed in the union, for it is our friend.

"If the union had won we would have been safe. Two of our demands were for adequate fire escapes and for open doors from the factories to the street. But the bosses defeated us and we didn't get the open doors or the better fire escapes. So our friends are dead."

15. PROTECTION

. . . man's peculiar vice.
—CANTO XI:25

ON THE DAY of the fire the Triangle Shirtwaist Company carried insurance totaling $199,750.

Months before the fire, the New York State Assembly had appointed a special committee to investigate corrupt practices among insurance companies other than life. The investigation was touched off by a scandal resulting from the accidental discovery that an insurance firm had bribed a number of state legislators.

After the fire, insurance expert Arthur E. McFarlane summarized in *Collier's* magazine the findings of the committee and offered Triangle as a classic example of a "rotten risk." He charged that if it was possible to obtain $100,000 worth of insurance on a $50,000 value, the purchaser would naturally tend to be careless and negligent. If a fire can mean gain instead of loss, he asked, why should even the most dangerous condition be corrected? "And if behind such an insurer we have an insurance system which permits and encourages over-insurance, which feels no obligation to inspect, remove dangers, or do anything whatever other than make insurance rates proportionate to the risk, the fire will follow almost as certainly as if kindled with matches and gasoline."

Factories such as Triangle were fire traps because of the power and the activities of the insurance monopoly known as the New York Fire Insurance Exchange. McFarlane charged that the exchange had driven "factory mutuals" out of New York. The mutual insurance companies would have provided genuine protection against fire losses and would have stimulated fire prevention activities. These mutual insurance companies had operated successfully for years in textile mills, especially around Fall River. They offered low-rate insurance in return for the installation of safety devices such as sprinklers.

In 1895, a number of factory and building owners in New York City experimented with the installation of sprinklers. They asked their insurance companies for commensurate rate reductions. When the stock insurance companies refused to grant the reductions, the owners made their own reciprocal arrangements and threatened to form their own insurance association. The rates were reduced by the companies.

In spite of this concession, the single factory owner still found the cost of installing sprinklers high. To meet this problem a group of ten companies selling insurance and employing engineers proposed a novel plan. They would make New York factories safer at no extra cost to their owners.

This was to be done by installing a sprinkler system, thus improving safety and reducing the risk of loss. For these improvements the owner would be entitled to a reduction of his insurance rates. But the plan was to continue to charge the owner the higher "unsprinklered" rate and to apply the difference toward paying off the cost of the sprinklers.

The plan proved popular. Unfortunately, it cut into the insurance agents' commissions and the brokers' fees.

The Asch building, for example, represented a total of $1,647,000 worth of insurance. At the "unsprinklered" rate, premiums for Asch and his eight tenants totaled about $15,000, of which the broker's share would be $1,600.

But a "sprinklered" New England or South Carolina cotton mill with the same amount of insurance paid about $1,100 in annual premiums. The agent or broker got about $60.

"A third of New York's insurance profits were being paid by its Asch buildings and Triangle factories," McFarlane pointed out. "If the property of the factory owner and the lives of his workers were made safe by this system of installing automatic sprinklers, the annual income of New York insurance agents and brokers might be diminished by at least $2,000,000."

The brokers and agents were the controlling power in the Exchange. They forced the withdrawal of the Exchange license to sell insurance from the ten companies sponsoring the deferred-cost plan. One powerful firm, I. Tanenbaum Son and Company, resisted the move by the Exchange; the others were compelled to give up the plan.

But despite the higher "unsprinklered" rates thus guaranteed, the insurance companies found business with shirtwaist companies unprofitable. They had too many fires. And there were reasons to believe there would be even more fires in the future. McFarlane cited as contributing causes for this the costly shirtwaist strike and the threatening challenge of the one-piece dress.

By the end of 1910, one large insurance company had canceled 693 out of 1,749 policies it provided for garment shops. In 1910, it had paid out in 42 shirtwaist fires. But in 1911, with only 1,056 factories covered, it paid out in 81 shirtwaist fires. The trade publication *Insurance Monitor* declared that such shirtwaist factory fires "were fairly saturated with moral hazard."

The insurance companies, confronted by this challenge, could take one of two possible courses of action. They could inspect and appraise the business conditions of all who had had fires and weed out all rotten risks. In this way, said McFarlane, there would be a concrete discouragement of fires and a positive incentive to prevent fires inasmuch as they would not be covered by indemnity.

Or the companies could take the second course which would be to estimate carefully the coming increase in the number of fires, raise their rates accordingly, let the fires come "and by keeping the rates up long after the crisis is past, profit by those fires."

The New York Fire Insurance Exchange began to study rate revisions in 1910. It decided to raise them at the start of 1911. "The rate committee, during the very month in which 146 were burned to death, even decided upon the price at which employees should be allowed to smoke in factories like Triangle," McFarlane wrote.

By June, 1911, increases totaling about 35 per cent were put into effect. "The payment for those deaths by fire in March had been arranged for in advance."

The fire on March 25 was not Triangle's first fire. In fact, the company was what insurers called a "repeater." This is the record:

April 5, 1902—5:18 A.M., ninth floor, Asch building. Cause unknown. Insurance collected: $19,142.

November 1, 1902—6:00 A.M., ninth floor, Asch building. Cause unknown. Insurance collected: $12,905.

November 10, 1904—6:57 P.M., dwelling of Isaac Harris, 845 West End Avenue. Cause, carelessness with matches. Loss, small. Insurance collected. Amount unknown.

April 7, 1905—11:00 P.M., factory of Triangle Waist Company, 49 West Third Street. Cause, unknown. Fire did not originate on premises. Insurance collected. Amount unascertainable.

April 12, 1907—1:15 A.M., Diamond Waist Company, owned and operated by Triangle at 165 Mercer Street. Insurance collected. Amount unascertainable.

In 1908 and 1909—exact dates unascertainable—two small fires, probably caused by smoking, occurred at Triangle. They were put out at once. No insurance claims were made.

April 27, 1910—7:18 P.M., at Diamond Waist Company, 187 Mercer Street. Cause unknown. $17,000 insurance carried. Amount collected unascertainable.

Did this record mean Triangle had had increasing difficulty maintaining its coverage or getting new policies? On the contrary, it carried its greatest amount of insurance coverage at the time of the March 25 tragedy.

How was this possible?

Triangle was represented by the powerful brokerage firm of Samuels, Cornwall and Stevens. No insurance company dared refuse coverage to this firm, says McFarlane. In any case, there was little risk for the individual insurance company. Total coverage of Triangle was by syndicate. This meant that many companies joined to cover the total $199,750. But each took only a small sliver.

McFarlane described the mechanics: "There is imprinted on the policy a small violet or blood-colored stamp reading 'Other Insurance Permitted,' and the broker goes down the line to get the next. No company is compelled to 'stay on the risk' after the second fire or the third." A company could choose the amount it would cover: $2,000 or $3,000 or $5,000.

There were almost 150 companies in New York selling this type of coverage. The broker could choose among them. Of the twenty-four companies which together had covered Harris and Blanck in 1902, the names of only five appeared among the thirty-seven companies that were liable after the 1911 fire.

The strike had hurt Triangle; new style developments were cutting into sales, and the firm had in fact begun production of the one-piece dress in an effort to hold its customers.

In January, 1910, the firm increased its insurance by nearly $100,000. In July, 1910, it refused to make a statement of its financial condition to a credit agency and as a consequence lost its credit rating. This was later restored "only by presenting a statement on which their whole resources were based, practically, upon the amount of insurance they carried."

In October and November, 1910, Triangle renewed $74,750 worth of insurance that had lapsed. In December, 1910, it increased its insurance by $25,000. "Before giving it, no company asked for any inventory or any evidence of added value. To have done so would have hurt that company's interests with the great brokerage firm of Samuels, Cornwall and Stevens," McFarlane commented.

But Triangle was hard pressed for money. On the morning of the day of the fire a provisional meeting of its creditors was held. The premium for $75,000 of insurance was not paid for until April 3, nine days after the fire.

McFarlane compares the Triangle shop to the "coffin ships" in the early history of British shipping. In the one case a fire and in the other a sinking, "could only be a business blessing, why should they do anything to prevent one?" The owners of the rotten hulks never actually did anything to sink them. But one old sailor had put it this way: "Her'll go down time enough, wi' the weight of her insurance, an' the things they ha' left undone."

The record indicates that with Harris and Blanck, April was the worst month. "If such things were a matter of interest to the stock fire insurance business," McFarlane insisted, "you would think that in the weeks preceding April they would have sent to see if there were no conditions discoverable which might in some way make for April fires. No one was sent."

But on February 11, 1911, an inspector from the New York Board of Fire Underwriters walked through the Triangle factory. At that time, nearly 900 dozen muslin and lawn garments were being made each week. Rags, cutaways, lint had been accumulating for almost a month since ragman Louis Levy's last visit.

As an agent of the Board, the inspector represented all of the insurance companies involved in covering Triangle, Asch, and the other seven tenants of the building. In theory, McFarlane stresses, this inspector was there to protect both insurer and insured. "There is still a wide, popular belief that the insurance business desires only to prevent fires. Here, then, was the very best chance to prevent one," he continued.

"The New York Board knew that as a 'repeater,' the Triangle Waist Company had long been a 'rotten risk.' The whole shirtwaist industry was then potentially a 'rotten risk.' Harris and Blanck had been for months financially a 'rotten risk.' Because of the lack of all fire protection, the Triangle factory was physically a 'rotten risk.' And now, by those ever-accumulating rags, this initial and inherent rottenness was plainly being, hour by hour, increased.

"That inspector must have seen those rags," McFarlane continued. "It would have been difficult not to! He did nothing! It was not within his province to do anything. The interest of his companies was guarded in the insurance rate. The moment a match head or cigarette butt dropped unheeded into one of those great rag bins, what must happen could be averted by no human power. Yet, smoking, too, was all a matter of insurance rates."

After the fire, Harris and Blanck retained the firm of Goldstein and Company as their public adjusters. By July 24, the company received a total of $8,500 for services rendered. McFarlane was unable to discover the nature of those services.

Formally, they were to determine the amount of indemnity due Triangle. But even though, admittedly, Triangle goods had been totally destroyed, the company had kept no inventory record which it was able to produce. It had first claimed that there had been such a record but that it had been destroyed in the fire. But later it admitted it had kept no such books, that it noted its inventories on slips of paper which were kept not in the safe but in an office desk that had been burned to ashes.

Nevertheless "proof of loss" submitted by the adjusters was quickly accepted. No account was taken in this "proof" of poor sales and overstock. On the contrary, the "proof" disregarded the injury the company had suffered from the strike and assumed company earnings were at a rate three times greater than before the walkout.

Only one of the thirty-seven insurance companies covering Triangle balked at paying its share of the total indemnity. The Royal Insurance Company asked for an investigation of the circumstances of the fire. Royal did not generally refuse to pay. In fact, even while the fire was still burning in San Francisco after the earthquake, Royal had bucked the large group of insurance companies that had offered immediate settlement of claims at the rate of 75 cents on the dollar to the stunned citizens of that city. Royal brought most of them back into honorable line by announcing it would pay 100 cents on the dollar.

Barrow, Wade, Cuthrie and Company, chartered accountants, were retained to examine whatever Triangle books and documents were still available. They found no reason to believe that the stock destroyed was worth more than $134,-075, at the most.

Nevertheless, George R. Branson, head of the adjusting committee employed to protect the interests of the companies and the public in the settlement, wrote to Max Steuer, attorney for Harris and Blanck, assuring him that there would be little difficulty in the matter. "And though there was blood on every policy," McFarlane concluded, "the insurance companies, with the exception of the Royal, paid the loss practically in full as demanded."

The proprietors of the Triangle Shirtwaist Company collected in indemnity $64,925 in excess of any claim for which they could furnish legal or convincing proof of loss, according to McFarlane.

The rate: $445 clear per Triangle worker killed.

16. JUSTICE

*Such sin unto such punishment condemns **him.***
—CANTO XVIII:95

ON APRIL 10, at the request of Assistant District Attorney J. Robert Rubin, Samuel Bernstein and Louis Brown were ejected from the Criminal Courts building where the Grand Jury was hearing witnesses. The *Times* declared, "it had been intimated that they were attempting to tamper with Grand Jury witnesses." Italian Consul Giacommo Farabonni, to whom many of the young Italian workers had turned for help after the fire, reported that seven girls had made affidavits declaring they had been approached.

The next day the indictment of Harris and Blanck was handed up to Judge O'Sullivan in the Court of General Sessions.

But it was more than eight months later—on December 4—that the trial began with the selection of a jury. Judge Thomas C. T. Crain, small-featured, gray-haired dean of the General Sessions bench, presided. As their counsel, Harris and Blanck took the noted lawyer Max D. Steuer, who, although only forty-one years old, already enjoyed a towering reputation for shrewdness and resourcefulness in the court room. The people were represented by Assistant District Attorneys Rubin and Charles S. Bostwick.

Eight months after the fire, public indignation had blunted considerably. Only in the hearts of the families of the dead did the sense of loss continue to cut with undulled sharpness. But even for them the demands of daily living had forced adjustments to grief and loss, and for many the single consolation was the vague hope that their longing for justice would somehow be fulfilled.

By the end of the first day six jurors were selected. Steuer had challenged seven talesmen, the prosecution ten. Steuer asked each talesman if he was in any way connected with a labor union, whether he read the New York *Call* or any other paper that had strongly condemned the Triangle partners. He inquired from each if he would be influenced by the presence in the court and the corridors of people dressed in mourning who might even burst into tears.

The courtroom and the corridors were, in fact, crowded by friends and relatives of Triangle victims. They made the first days of the trial a nightmare for Blanck and for Harris. On Tuesday morning, December 5, as the two partners stepped out of the elevator on the second floor of the Criminal Courts building with Steuer, a shrieking mob of women who had been waiting in the corridor came running toward them.

"Murderers! murderers!" the women screamed. "Make them suffer for killing our children!"

Policemen and court attendants rushed to the rescue. Mad with sorrow, the women fought them off. Then the officers surrounded Harris and Blanck and elbowed them into a smaller corridor leading directly to the courtroom.

Three women fought their way through and when they reached the defendants pushed photographs up close to their faces and yelled: "These are our children! Give us back our children!"

Throughout the morning, the crowd grew larger. When, during a recess, Bostwick and Rubin came into the corridor, they were immediately surrounded. Shouts rose in several languages. When the two partners also emerged, the crowd again became menacing and the hall had to be cleared.

The following day a full police guard patrolled the court house. The corridors were kept cleared. But when Harris,

Blanck, and Steuer left the building during the lunch recess, they were spotted walking toward Broadway. A crowd began to follow them, growing as it moved closer. Detectives came to the rescue by pushing the three men into a restaurant on Franklin Street and then standing guard in the doorway. The ordeal was repeated as the three made their way back to the court building.

The jury consisted of Leo Abraham, real estate; Anton Scheuerman, cigars; William E. Ryan, salesman; Harry R. Roeder, painter; Morris Baum, salesman; Charles Vetter, buyer; Abraham Wechsler, secretary; Joseph L. Jacobson, salesman; William O. Akerstrom, clerk; Arlington S. Boyce, superintendent; Victor Steinman, shirts; H. Houston Hierst, importer.

In opening, Bostwick's task seemed simpler than the one confronting the defense. Gently and courteously, he announced he would question his witnesses who would set the scene of the tragedy and establish the physical and temporal framework of the events. Thereafter, he would call those who might substantiate the alleged fact that a door on the ninth floor, through which many who had perished could have escaped, was locked at the time of the fire. More particularly, he would then show that because of the illegally locked door, Margaret Schwartz, in whose horrible death was gathered up the deaths of her co-workers for the purpose of rendering justice, had been unable to escape and had perished in the flames.

The trial was to last for little more than three weeks. But in its eighteen working days, 155 witnesses were called. The people called 103; the defense 52.

Steuer came from the same East Side whose slums and tenements were the living quarters of most of the witnesses called by the people. He knew them well, spoke their language, sensed their fears and their resentments, and his first cases in the practice of law involved broken contracts, unpaid bills, and other ethical and legal dislocations in those lower depths. All this gave him many advantages over Bostwick.

As the trial progressed, the shop girls, dressed in their

holiday best, followed one another to the witness chair. Awed by the majesty of the law, sometimes adding fright to the inexperience of youth, often struggling with a language not always clear even in translation, they replied in low tones until the moment when memory rekindled the horror.

Bostwick was their soft, sympathetic champion; Steuer, their dangerously prodding inquisitor.

At times, Steuer seemed to enjoy the contest. When sixteen-year-old Ethel Monick refused to be cowed, he said to her, "You do like to argue some, don't you, little girl?"

He asked her why she had never spoken to Mr. Harris about the Washington Place door. He was not satisfied with her answer— "I used to be afraid of him"—and registered disbelief. Whereupon Ethel Monick added she hadn't exactly meant afraid, "But, you know, I was like nothing to them because I was only a working girl."

He asked Celia Walker to describe how she had made her way across the shop, and she told the court she jumped across the aisles from machine table to machine table. Steuer questioned, "Was your skirt about as tight as the skirt you've got on now?"

The young lady answered, "No," but added that she used to practice jumping. A stickler for detail, Steuer queried, "Did you find out how broad a jump you could make?"

For a time, as burning death had come closer, Kate Gartman had considered leaping from a ninth-floor window and in that time, she told the court, "I was cool and nice and calm."

"Sure, now when you are nice and cool and calm that way, you got to the window?" Steuer questioned.

"I was as cool and calm as you are here," Kate Gartman replied.

"Now, I'm not half as cool as you think I am," was Steuer's comeback.

Kate Gartman's forwardness was less the result of courage than of inability to comprehend. She ended the spirited exchange by stating: "I could not understand what you meant by talking because I don't think I am so excellent in the English language to understand it all. Therefore you have to excuse me if I made a mistake."

Bostwick described the witnesses he called as "of tender years—most of them not able to speak the language, not of great intelligence . . . working at their machines, working and working with no time to look up."

But for Steuer they were merely people whose motives and honesty he suspected, "people who came here—most of them with law suits. Many of them, because they lost their dearest relatives, are not telling the truth."

Except for technicians and members of the Fire Department, the people's side called only Triangle employees to the stand. On the other hand, of those called by the defense only a half-dozen were production workers with no family connection with one or the other partner. The rest were either supervisory or nonproduction employees, salesmen, buyers, or others who entered the premises from time to time but did not work at Triangle.

Throughout the trial and especially in his summation to the jury, Steuer continued to voice doubt about the ability, and even the willingness, of the people's witnesses to tell the truth. At the same time, doubtless with the composition of the jury in mind, he made frequent references to the salesmen who had appeared as defense witnesses. Could such upstanding persons, he asked several times, commit perjury?

The defense argued that the great loss of life was due to the failure of the workers on the ninth floor to turn to the Greene Street exit. It claimed the Washington Place door on that floor was actually open; however, escape by way of the staircase on that side had been impossible because the fire penetrated into the stairwell.

Why had the girls on the ninth floor jammed up in front of the Washington Place elevators, Steuer asked? Why had they acted against all previous practice? Why had they not gone to the only exit they claimed was the one they were allowed to use? Why, instead, had they ended up in front of a door which they themselves said they had never used before, which they themselves said they had always considered to be locked, and which, at the tragic moment, they said they found to be locked? Steuer pressed these questions with vigor, attempting to show that the girls knew the best and proper way of making a quick exit from the ninth floor, but that moved by fear and panic they were trapped by their own irrational behavior.

Bostwick, in turn, argued that the fire, which had started on the cutting tables at the Greene Street side of the eighth floor, had first lapped in through the windows of the ninth floor on the same Greene Street side and had soon cut off escape through the Greene Street door. In contrast, there were no windows near the doors on the west wall and therefore there could not have been fire in the Washington Place stairwell, which, he insisted, could not be used by the ninth-floor girls only because the door was locked.

If there had been fire in the Washington Place stairwell, Bostwick continued, it would have had to come out of the eighth-floor door early in the emergency. Eventually, the fire did burn through. But not before the eighth-floor girls had come through the doorway. Nor had there been any suction to pull the eighth-floor flames up the staircase even if they were in the stairwell; the ninth- and tenth-floor doors were closed, and the skylight on the roof over the well was unbroken.

It was not true, said Bostwick, that all the girls had immediately made a dash for the Washington Place side. Some had escaped through the Greene Street door. Others were diverted from doing so. Rose Mayer saw girls running toward the Washington Place side, so "I ran toward there, too," she said. Somehow, Ethel Monick had thought she heard the elevators on the Greene Street side crash, so she turned in the opposite direction. But others had run across the shop because they had seen flames in the Greene Street stairwell. Some were trapped because they had planned to use the Greene Street exit but had first gone to the dressing room to get their coats and hats.

The defense in turn relied heavily on the fact that State Labor Department inspectors had never reported finding the Triangle doors locked. The Commissioner of Labor, called as a defense witness, "swore before you," Steuer stressed in summing up, "that no notice is given to the people whose factories they are about to inspect." Which, he asked, were to be believed: those still suffering the hurt and the loss caused by the fire, or this "disinterested person working for the State of New York whose duty it is to see that the law is absolutely complied with?"

But State Labor Commissioner Williams actually provided little help for the defense. Under cross-questioning by Bostwick, he was unable to read correctly a report by one of his inspectors. The report form contained one question about doors: "Open out practical?" The inspector had put a pencil check above and between the words "out" and "practical." Did this mean that particular door opened outward? Or did it mean the door opened inward but that it was practical to make it open outward? The Commissioner couldn't say.

Bostwick then pointed out that the inspector had to note on each report filed the identity of the "person in authority" who was questioned during the inspection. Commissioner Williams insisted this had to be a person of superior position in the firm. Bostwick then showed him an inspection report on Triangle. The report had been accepted as proper. The "person in authority" was named Edna Barry.

Did the Commissioner know what superior position she held? He did not. She was, said Bostwick, one most suitably situated to signal the arrival of an inspector throughout the factory, for she was the Triangle telephone switchboard operator who had been absent on the day of the fire.

Midway through the trial, Bostwick had a large package brought into the courtroom. When it was unwrapped, it turned out to be a velvet-lined box with a framed-glass lid, a large container of the kind in which table silverware is generally kept. He had kept it locked in one of his own rooms at home, and his children had been warned to resist curiosity and to keep away from the box. In that box was the lock of the Washington Place staircase door, still attached to a piece of wood panel, its bolt still sticking out.

Two workmen, Giuseppi Saveno and Pietro Torchin, cleaning the rubble on the ninth floor, had found it in the debris on April 10. Detective Francis Flynn had seen them pick it up from a pile of rubbish about 11½ feet from the doorway. Unfortunately, he had made no memorandum of this in his notebook, at that time.

Steuer fought bitterly against having the court receive the exhibit in evidence. In an angry onslaught he addressed the court:

"On March 25th a fire occurred and on March 26th all the conscience of the city is stirred by the terrible catastrophe. And on the 27th of March the theory of the locked door is already made public. Hundreds upon hundreds of people go into the debris and seek the bodies. The Fire Department makes a conclusive and minute and detailed search into the debris. The whole question that is being agitated in the press day after day is locks, locks, locks. Nothing is found.

"Then on the 10th of April, as from a clear sky, a detective goes to the premises and within 25 minutes a lock is discovered. Fifteen days after the fire they seek to say that was the lock of the door.

"Why argue this? Where is the lock of the tenth floor where the fire did not do damage? Where is the lock of the sixth floor where the fire did not touch it? A most mysterious disappearance.

"Right after the discovery by the newspapers of this lock, there is no lock on the sixth floor. All of the locks, may it please your Honor, in that building from the tenth floor down to the basement were identical. No lock on the eighth floor, your Honor, has been discovered. Why, I ask, is there no lock for the tenth floor?

"But the one lock that is wanted—that is found on the 10th day of April. And they say that exclusive opportunity has nothing to do with this case."

But Bostwick was undismayed. Patiently, he set out to

prove that this and only this could possibly be the lock that trapped the girls on the ninth floor.

First he pointed out that this was a unique lock, unlike any other that might have been found on the ninth floor. It was a mortised lock, one set into the wood so that its sides were flush with the wooden surface of the door instead of being set onto the outside surface of the door. No other doors on the floor could have had mortised locks. The dressing room door was closed with padlocks. The toilet doors had no locks. A sliding door in the partition at the Washington Place side was only ⅞ inch thick—too thin to hold a mortised lock.

Secondly, Bostwick continued, this lock was different from all other locks in the building. He began the chain of proof for this by calling to the stand Battalion Chief Thomas Larkin who described the fire that had occurred on the ninth floor on November 1, 1902. That was when a new lock, the one in the box, was put into the door.

Then Bostwick proceeded to tie the lock in with the 1902 fire. He called Charles W. Baxter, superintendent of the J. W. Clark Co. which had repaired the ninth-floor door in 1902. Then he called Emil Woehr, a dealer in builder's hardware who had purchased the lock along with other materials from the Reading Hardware Co. and had sold it to J. W. Clark Co. in December, 1902. Finally, he called Francis J. Kelly, who had worked for Reading and had filled and delivered the order to Woehr Brothers.

The chain was complete, and Bostwick sought to strengthen it with the word of an expert. He called John D. Moore, a former state commissioner now working as a consultant engineer. Moore swore that the lock could not have come from the Greene Street door because that was a right-handed door and the lock in the box was a left-handed lock which could come only from a left-handed door. The Washington Place staircase had a left-handed door.

Steuer chewed on it. His great talent for detail, his informed imagination came into play. He demanded to know what made a door left-handed.

It was a door which, in being opened, moved counterclockwise, Moore replied.

And what was a left-handed lock?

Moore carefully explained that the extended bolt or tongue of a lock can be drawn back into the lock when closing the door in one of two ways. One way is by turning the knob of the lock. The other is to push the door closed until the beveled side of the bolt hits the metal lip or strike attached to the doorframe. Because it is beveled, the bolt will then gradually be forced back into the lock, ready to spring out again when it is finally in line.

A left-handed lock, Moore continued, is one that will do this on a left-handed door. A right-handed lock placed on such a door would, at the moment of strike, have its beveled face on the wrong side and would present a flat surface to the metal lip, thus stopping the door from closing.

The lock in the box was a mortised lock, a left-handed lock, and the Washington Place door was a left-handed door, Moore concluded.

Now that Steuer understood the difference, he began to use that understanding to pick apart the lock. He had Francis Kelly back on the stand. Kelly had told of handling locks for twenty-one years and had positively identified the left-handed lock in the box as one that he had handled.

Steuer handed Kelly a right-handed lock. He asked Kelly if he could take it apart and convert it into a left-handed lock.

Kelly answered he could, by reversing the latch bolt.

Could any lock be made right-handed or left-handed?

Yes.

When the locks leave the factory, are they different?

No, we ship them all right-handed.

When you make locks, are they made left-handed or right-handed?

We build them so that they can be reversed.

Then there isn't any such thing as a right-handed or left-handed lock, Steuer concluded disdainfully, turning away from the witness.

Its usefulness thus undermined, People's Exhibit No. 30 was received in evidence.

The lock had a key.

In opening for the defense, Steuer told the jury: "Louis Brown will tell you that year in and year out, on the Washington Place side on the eighth and ninth and tenth floors, the key was all the time in the lock." He gave no explanation why this was the case where doors were alleged to have been unlocked.

Brown, the heroic machinist who finally got the eighth-floor Washington Place door open, had apparently taken for granted, in the moment of his heroism, that the door was locked.

"I naturally thought that they must have locked the door," he told the court. "There was a key always sticking in that door. . . . All I tried to do was to turn the key in the lock. But the key wouldn't turn to unlock the door. It did not turn. So I pulled the door open. . . ."

Steuer explained to the jury that "the key would not turn because the door was not locked." But he did not explain the need for an ever-present key if the door was kept unlocked during working hours.

Only defense witnesses could describe the key.

May Levantini said it hung by a piece of colored string. Ida Mittleman testified it was attached to the door by a piece of tape, "or something," which William Harris said was a length of white lawn. And cleanup man Nathan Salub told the jury that "once at the door I saw the string was worn out and I myself picked up a piece of white goods, a strong string, and I tied it to the door."

Levantini said the string was ½ inch or more in width; Harris estimated a width of 2 to 3 inches. Levantini described the string as being ½ yard long; building engineer Casey called it 6 to 8 inches long and insisted that the key was also 6 inches long.

Not one witness for the people from the ninth floor mentioned the key or the string. But the defense called two witnesses who had not only seen the key but also had actually used it.

Steuer had some difficulty in reviewing the testimony of Gussie Rapp, a ninth-floor forelady and one of the two witnesses. He declared to the jury:

"Gussie Rapp told you that the Washington Place door was always unlocked and that the key was in the door. I want to withdraw that statement. What she did say was she said that the door was always unlocked—no, that was not it. She said this: That there may have been—she has no positive recollection of it—that there may have been a time when that door was locked when she came to it. But if it was, she simply had to turn the key and pass through."

The other witness who handled the key, even on the day of the fire, was May Levantini who declared: "The door was locked. I turned the key that was in the lock and I opened the door."

Steuer called more than a score of witnesses who testified to the ease and the frequency with which people ordinarily passed through the ninth-floor Washington Place staircase door. He called seven salesmen, five foreladies and assistants, three porters, two shipping clerks from the tenth floor, two watchmen of whom one doubled as a quitting-time purse inspector, the company superintendent, the machinist, the painter, a big department store buyer, the ragman, the building superintendent, the building engineers, and others.

In turn, the people called more than a dozen workers who swore that in the moment of peril they had tried to exit through that door and had found it locked.

Joseph Brenman had pushed through the crowd: "I tried the door, I took hold of the handle and pulled it." It wouldn't open. Anna Gullo said: "I tried the door. The door was locked." Mary Bucelli declared: "I tried to open the door but I couldn't." Ida Nelson swore: "I pushed at the door but I could not open it."

At Bostwick's request, Katie Weiner, who had lost a sister

in the fire, rose from the witness chair in a hushed courtroom and stepped to a side door within easy view of the jurors. She put both hands on the door knob and began to wrestle with it, her voice filling with tears as she said:

"I pushed it toward myself and I couldn't open it and then I pushed it outward and it wouldn't go. I was crying, 'Girls, help me!' "

Ethel Monick insisted she was the first to reach the door. "I tried the door and I could not open it. So I hollered, 'Girls, here is a door!' and they all rushed over and tried to open it. But it was locked and they hollered, 'The door is locked and we can't open it!' "

In the confusion, many had thought they were the first to reach the door.

"I was just about the first one to catch hold of the knob and I twisted it and turned it and it was closed and it wouldn't open," Rose Mayer swore.

"I was the first at the door," said Rose Glantz.

"I was the first one at the door," said May Levantini.

May Levantini, mother of three, had saved herself by sliding down the elevator cable from the ninth floor. She was the strongest witness for the defense, for in her words was the proof that those who died at the Washington Place door were not the victims of a locked exit but of fire in the staircase.

She had run to the door and found "the key was right in the door tied to a string. I turned the key. I opened the door. I looked out and I saw the girls running down from the eighth floor. And as I looked over that way, flames and smoke came up and they made me turn in. I turned right in and ran to the elevators."

In support of Levantini's testimony, Steuer called the Mittleman sisters, Ida and Anna. He reminded the jury that Anna had said she followed Levantini through the open door and looked down over the banisters. "The door was open and she saw the flame and she saw the smoke and she saw the girls going down the stairs. And fearing that she could not get down safely, she turned back through the open door."

Mary Alter, the tenth-floor typist, strengthened the argument by telling how she too had opened a door to the Washington Place stairs, the door on the tenth floor, and had seen the smoke and the flame below. When Bostwick asked if she was certain she had seen flames she replied that "as I looked over the rail I saw a great volume of smoke and I saw a red streak."

Certainly if Mary Alter had seen the flame from the tenth floor, May Levantini and Anna Mittleman had seen it from the ninth, Steuer inferred.

But Bostwick wondered out loud how there could be both flames and living people in that eighth-floor doorway at the same time. "Every girl went through that door and still there was no flame," he stressed. Machinist Brown had been the last one out that door and was untouched by fire. No injured were found in the stairwell.

Bostwick was also bothered by the mutual dependence of the Levantini and Mittleman testimony. He read from the record to show that Ida Mittleman admitted her sister had refreshed her memory about the incident at the door. "She told me about how the two of us did go in the hall and that is what I do remember; and about the door, she said she saw me open it," Ida Mittleman had declared.

In turn, Anna Mittleman had admitted discussing the events with May Levantini. "She told me she opened the door," Anna Mittleman had said.

Bostwick insisted that "Levantini lied on the stand," that "it was not until after she had seen Flecher and Bernstein that she said she opened the door."

Bostwick added it up for the jury: "Anna tells May and May tells Anna, and all of this is after they have seen Bernstein. But they don't remember anything about it. They simply are convinced from what they have heard that Levantini was telling them the truth. Anna says that Ida told her so and vice versa."

Just as May Levantini was completing her direct testimony, Juror No. 3 directed a question to her. He wanted to know:

"What did you do to the door after you came back in?"
"I pushed it in back of me and ran for the elevator."
"You closed it?" asked the juror.
"Yes sir."
"Did you lock the door after you came back in?"
"No, sir. I would never think of turning that key again."
Yet no others got through the door.

Because it was locked, the people charged, it was at this door that Margaret Schwartz had died.

Bostwick relied heavily on Kate Gartman and Kate Alterman to prove this charge. These two had emerged with Margaret Schwartz from the dressing room.

"In going to the elevator doors, there is a partition," Kate Gartman said. "I had seen Margaret Schwartz grab hold of the partition. I don't know—she kind of leaned or fell toward the dressing room. Then I saw the elevator come up."

Bostwick saved Kate Alterman for the last. And it was as if Steuer had been waiting for her. Earlier he listened to Lena Yaller tell how she had escaped from the ninth floor to the roof. He asked her to tell the story again. When she finished, Steuer asked if she could tell those words over again. Apparently convinced that Steuer was questioning the accuracy of her account, she quickly answered: "I could tell them ten times."

Now he listened to Kate Alterman's direct testimony of Margaret Schwartz's last moments:

> Then I went to the toilet room. Margaret disappeared from me and I wanted to go up Greene Street side, but the whole door was in flames, so I went and hid myself in the toilet rooms and bent my face over the sink, and then I ran to the Washington side elevator, but there was a big crowd and I couldn't pass through there. Then I noticed someone, a whole crowd around the door and I saw Bernstein, the manager's brother, trying to open the door, and there was Margaret near him. Bernstein tried the door, he couldn't open it.
>
> And then Margaret began to open the door. I take

her on one side—I pushed her on the side and I said, "Wait, I will open that door." I tried, pulled the handle in and out, all ways and I couldn't open it. She pushed me on the other side, got hold of the handle and then she tried. And then I saw her bending down on her knees, and her hair was loose, and the trail of her dress was a little far from her, and then a big smoke came and I couldn't see.

I just know it was Margaret, and I said, "Margaret," and she didn't reply. I left Margaret, I turned my head on the side and I noticed the trail of her dress and the ends of her hair begin to burn. Then I ran in, in a small dressing room that was on the Washington side, there was a big crowd and I went out from there, stood in the center of the room, between the machines and between the examining tables.

I noticed afterwards on the other side, near the Washington side windows, Bernstein, the manager's brother throwing around like a wildcat at the window, and he was chasing his head out of the window, and pull himself back in—he wanted to jump, I suppose, but he was afraid. And then I saw the flames cover him. I noticed on the Greene Street side someone else fell down on the floor and the flames cover him.

And then I stood in the center of the room, and I just turned my coat on the left side with the fur to my face, the lining on the outside, got hold of a bunch of dresses that was lying on the examining table not burned yet, covered my head and tried to run through the flames on the Greene Street side. The whole door was a red curtain of fire, but a young lady came and she wouldn't let me in. I kicked her with my foot and I don't know what became of her.

I ran out through the Greene Street side door, right through the flames on to the roof.

In cross-examination, Steuer took the long way round, asking numerous questions about the witness's family, her two weeks' wait to testify, and her visit to the Asch building with Bostwick and Rubin. Then suddenly, he called on her to repeat her account of the death of Margaret Schwartz. Kate Alterman told it for the second time in these words:

> I went out from the dressing room, went to the Waverly side windows to look for fire escapes, I didn't find any and Margaret Schwartz was with me, afterwards she disappeared. I turned away to get to Greene Street side, but she disappeared, she disappeared from me. I went to the toilet rooms, bent my face over the sink, and then went to the Washington side to the elevators, but there was a big crowd, and I saw a crowd around the door, trying to open the door; there I saw Bernstein, the manager's brother, trying to open the door but he couldn't.
>
> He left; and Margaret was there, too, and she tried to open the door and she could not. I pushed her on a side. I tried to open the door, and I couldn't and then she pushed me on the side and she said, "I will open the door," and she tried to open the door, and then a big smoke came and Margaret Schwartz I saw bending down on her knees, her hair was loose and her dress was on the floor and a little far from her.
>
> And then she screamed at the top of her voice, "Open the door! Fire! I am lost, there is fire!" and I went away from Margaret. I left, stood in the middle of the room, went in the middle of the room, between the machines and examining tables, and then I went in.
>
> I saw Bernstein, the manager's brother, throwing around the windows, putting his head from the window —he wanted to jump, I suppose, but he was afraid— he drawed himself back, and then I saw the flames cover him, and some other man on the Greene Street side, the flames covered him, too.
>
> And then I turned my coat on the wrong side and put it on my head with the fur to my face, the lining on the outside, and I got hold of a bunch of dresses and

> covered the top of my head. I just got ready to go and somebody came and began to chase me back, pulling my dress back, and I kicked her with the foot and she disappeared.
>
> I tried to make my escape. I had a pocketbook with me, and that pocketbook began to burn, I pressed it to my heart to extinguish the fire, and I made my escape right through the flames—the whole door was a flame right to the roof.

Were these the words of a witness who had been coached and drilled?

The accounts were similar but not identical. The sequence of events was parallel. Much of the language was identical. But there were also important differences. The second account was shorter. It also mentioned the Waverly side fire escape, and Margaret's screams, which are missing in the first account. In her first telling Kate declared she said she would open the door but in the second telling she quotes Margaret as saying this.

Steuer asked wasn't the door "like a wall of flame?" "Like a red curtain," Kate answered.

"Now there was something that you left out," Steuer continued. "When Bernstein was jumping around, do you remember what that was like? Like a wildcat, wasn't it?"

"Like a wildcat," Kate Alterman replied.

"You did leave that out didn't you, just now, when you told us about Bernstein, that he jumped around like a wildcat?"

"Well, I didn't imagine whether a wildcat or a wild dog; I just speak to imagine just exactly," she answered.

Now Steuer switched back to details of the witness's life and the location of her machine in the shop. He had prepared the destruction of her acceptability.

Then he asked her to tell her story for a third time. This time she didn't forget the two points on which Steuer had questioned her after the second account. Also, this time she quoted neither herself nor Margaret as declaring "I will open the door," and in this third telling, there are no terrible screams from Margaret. Instead, the account is considerably shorter:

> I went out to the Waverly side windows to look for fire escapes. Margaret Schwartz was with me, and then Margaret disappeared. I called her to Greene Street, she disappeared and I went into the toilet room, went out, bent my face over the sink, and then I wanted to go to the Washington side, to the elevator. I saw there a big crowd, I couldn't push through. I saw around the Washington side door a whole lot of people standing, I pushed through and there I saw Bernstein, the manager's brother, trying to open the door; he could not and he left.
>
> Margaret Schwartz was there, she tried to open the door and she could not. I pushed Margaret on the side, tried to open the door, I could not and then Margaret pushed me on the other side, and she tried to open the door. But smoke came and Margaret bent on her knees; her trail was a little far from her and her hair was loose, and I saw the ends of her dress and the ends of her hair begin to burn.
>
> I went into the small dressing room, there was a big crowd, and I tried—I stood there and I went out right away, pushed through and went out and then I stood in the center of the room between the examining tables and the machines.
>
> Then I noticed the Washington side windows—Bernstein, the manager's brother, trying to jump from the window, he stuck his head out—he wanted to jump, I suppose, but he was afraid—then he would draw himself back, then I saw the flames cover him. He jumped like a wildcat on the walls.
>
> And then I stood, took my coat, turning the fur to my head, the lining to the outside, got hold of a bunch of dresses that was lying on the table, and covered it up over my head, and I just wanted to go and some lady

> came and she began to pull the back of my dress; I kicked her with the foot and I don't know where she got to.
>
> And then I had a purse with me and that purse began to burn, I pressed it to my heart to extinguish the fire. The whole door was a flame, it was a red curtain of fire, and I went right on to the roof.

There it was. Steuer had asked for them and her own lawyer had made no objection. So there was the "curtain" of fire and there was poor Bernstein's doomed brother jumping "like a wildcat." She had remembered.

Steuer could have asked Kate to repeat the story again. But Margaret Schwartz died no more that morning. The court adjourned for lunch.

After the recess, after he had elicited the fact that Kate had five sisters and her statement that only once had she tried to talk about the tragedy to them and they had stopped her, Steuer took her back to the point at which she stood in the center of the floor, between the machines and the examining table.

"Now tell us from there what you did," he commanded. Kate Alterman said:

> I saw Bernstein on the Washington side, Bernstein's brother, throw around like a wildcat; he wanted to jump, I suppose, but he was afraid. And then he drawed himself back and the flames covered him up.
>
> And I took my coat, turned it on the wrong side with the fur to my face and the lining on the outside, got hold of a bunch of dresses from the examining table, covered up my head, and I wanted to run, and then a lady came and she began to pull my dress back, she wanted to pull me back, and I kicked her with my foot —I don't know where she got to.
>
> And I ran out through the Greene Street side door, which was in flames; it was a red curtain of fire on that door; to the roof.

Once again, she had remembered the "curtain" of fire, the desperate man thrashing about "like a wildcat."

Steuer nailed it down. "You never studied those words, did you?" he asked.

"No, sir."

Bostwick rushed in to save his key witness. Kate Alterman could certainly give the account in other language. She had used different words to tell him about Margaret Schwartz's death the previous Saturday, Bostwick insisted.

"State to the jury why you tried to repeat the last time what you told Mr. Steuer in the same language that you used the first time you told Mr. Steuer."

"Because," Kate Alterman replied, "he asked me the very same story over and over, and I tried to tell him the same thing, because he asked me the very same thing over and over."

But irreparable damage had been done. Steuer asked if she could tell the jury the same words she had used in her written statement which she had made for Bostwick.

"Probably I can. My written statement was nine months ago," she answered.

"Tell us the words in the statement, please, now."

"Shall I tell you just as in the statement?"

"Yes, the words in the statement."

"Well, I gave a very long statement, I believe, to Mr. Rubin."

"Now, start with the words in the statement, please, and not an explanation, Miss, if you can. Tell us just how you started the statement, and then give us the words that are in the statement."

"Well, it would be 4:45 on Saturday, I think that I started this way, I am not quite sure, I don't remember just how I started the beginning of the statement, I can't do it to you."

"Mr. Bostwick asked you before whether you could tell again in the same words of the statement and you said yes. Now I suppose you did not understand the question that way, did you?" Steuer wound up.

"No, sir, I did not," Kate Alterman finished.

Was this the point at which Steuer won his case? Without making a charge, he had sought to demonstrate that the witness had been coached. But if that were so, would not her own attorney have cautioned against repeating identical language? Steuer, in a remarkable courtroom performance, had broken through the curtain of public sympathy for the shop girls and without once saying a word against Kate Alterman had yet impugned her reliability. Were the similarities in her four accounts more significant than the differences? Had she undermined herself through a fatal misunderstanding of what was wanted from her on the witness stand?

In charging the jury before it retired to seek a verdict, Judge Crain read Article 6, Section 80, of the Labor Law: "All doors leading in or to any such factory shall be so constructed as to open outwardly where practicable, and shall not be locked, bolted, or fastened during working hours."

When he carefully emphasized that "all" meant "every such door," Steuer took an exception.

Judge Crain then considered the meaning of "working hours." He told the jury the term included not merely the period of time during which persons are actually employed at their machines, cutting tables, examining tables, and desks, "but also a reasonable period of time for the exit of such persons from the place in which they are employed after the cessation of their work." Steuer again took an exception.

Then Judge Crain phrased the issue confronting the jury:

"You must be satisfied from the evidence, among other things, before you can find these defendants guilty of the crime of manslaughter in its first degree not merely that the door was locked—if it was locked—but that it was locked during the period mentioned under circumstances bringing knowledge of that fact to these defendants.

"But it is not sufficient that the evidence should establish that the door was locked—if it was locked—during such a period; nor yet that the defendants knew that it was locked during such a period—if it was locked.

"You must also be satisfied from the evidence beyond a

reasonable doubt that there was the relation of cause and effect between the locked door and the death of Margaret Schwartz.

"Was the door locked? If so, was it locked under circumstances importing knowledge on the part of these defendants that it was locked? If so, and Margaret Schwartz died because she was unable to pass through, would she have lived if the door had not been locked and she had obtained access to the Washington Place stairs and had either remained in the stairwell or gone down to the street or to another floor?"

The jury retired to deliberate at 2:55 P.M. on December 27. It polled itself three times.

On the first count, the vote was eight to two, with two abstentions.

On the second, it was ten to two.

Afterwards, juror Victor Steinman, a shirt manufacturer, declared: "I know I have not done my duty toward the people. But let me tell you I could not do my duty toward them and obey the Judge's charge at the same time, although I did vote once for conviction.

"I believed that the door was locked at the time of the fire. But we couldn't find them guilty unless we believed they knew the door was locked. It would have been much easier for me if the state inspectors instead of Harris and Blanck had been on trial. There would have been no doubt in my mind then as to how to vote."

On the other hand, juror H. Houston Hierst, importer, said his conscience was "perfectly at rest. I cannot see that anyone was responsible for the disaster. It seems to me to have been an act of the Almighty. I think that the Harris and Blanck factory was well managed. I paid great attention to the witnesses while they were on the stand. I think the girls who worked there were not as intelligent as those in other walks of life and were therefore the more susceptible to panic."

At 4:45 P.M., after deliberating one hour and fifty minutes, the jury brought in a verdict of not guilty.

The two partners received the verdict composedly. They went from the courtroom to an anteroom where Mrs. Blanck threw her arms around her husband's neck, sobbing with joy.

The two men tried to leave the courthouse as unobtrusively as possible. They were led through Judge Crain's private chambers and out through a door opening on a far corridor. But some of the crowd had managed to seep back into the hallways.

"We waited," Josephine Nicolosi remembers. "Then they came out. The people began to scream and cry, 'Give me back my daughter! Give me back my son! Justice! Where is Justice?' All the doors on that long hallway suddenly opened up and the judges stuck their heads out from their private chambers and they stood there in their black robes in the doorways, looking."

The acquitted men hurried down a flight of stairs to the ground floor, surrounded now by five uniformed policemen and two detectives, their friends and relatives trailing after. They were rushed through a pen filled with prisoners who had been found guilty in the Magistrate's Courts. They left the building through an exit on Leonard Street.

About 150 women, children, and men were waiting for them. For a minute there was silence. Then David Weiner, Katie's brother, whose seventeen-year-old sister Rose had perished in the fire, broke from the crowd, ran directly to the two partners and, shaking his fist at them, cried shrilly and mockingly:

"Not guilty? Not guilty? It was murder! Murder!"

Others took up the cry.

The two partners ran to the nearest subway station at Lafayette Street. David Weiner led the crowd in pursuit. The *World* reported that he collapsed on the steps leading down to the subway station and kept shouting hysterically: "Not guilty? Murder! Murder!"

Justice, in the best way men know how to find it, had been rendered. The jury had been given the impossible task of determining whether Harris and Blanck knew the door was locked at a specific time.

There could be no doubt that the Washington Place staircase door was at times locked during the working day. That certainty was not due so much to the lock that Bostwick introduced as to the fact stressed by the defense itself that the key was at all times attached to the door. Its own best witness, May Levantini, admitted having to turn the key in the lock in order to open the door when she went out onto the landing.

Could Margaret Schwartz have survived had she been able to pass through that door? It seems certain that having been able to pass through it she could then have been able to descend to the eighth floor and further. No one who had entered that staircase from the eighth floor had suffered burns. It was only after the fire had progressed considerably that the eighth-floor door may have become impassable.

Was the door locked at the moment Margaret Schwartz sank down into the smoke that asphyxiated her?

It seems certain that if there were times when that ninth-floor door was kept unlocked, those periods occurred only during the workday when there wasn't the slightest probability that a production worker or union member would go through it but when others, especially supervisors and foreladies, kept a steady traffic going between floors.

But when quitting time approached, when perhaps some worker would be tempted to take a quick detour down the Washington Place stairs, then the door was locked. For what other purpose was the key kept on a string tied to the door than to make certain that only those with authority to do so would dare to open it? That day, and on all other days, when work stopped the door was locked. There was only one way out for the workers: through the 30-inch opening in the Greene Street partition where the watchman stood to inspect pocketbooks.

Striving to impress the jury with the losses the firm had suffered through alleged pilfering by the girls in the shop, Steuer had brought into the court a large pocketbook of the kind that many of the workers carried.

"Just open it and see for yourself how big it is," he said and without warning tossed the bag to Assistant District

Attorney Rubin. Unsuspecting, Rubin opened the bag and drew out four shirtwaists. Bostwick immediately conceded that the pocketbooks were large enough for that purpose.

Steuer handed the four shirtwaists back to Blanck who caressingly folded each of the four waists and then piled them neatly one upon the other without any interruption of his attention to the proceedings.

Bostwick underlined the significance of the business with the pocketbook. "Never did the defendants themselves adduce such magnificent and convincing evidence to the jury as when Mr. Blanck, still thinking of the stealings, produced this bag in court with his shirtwaists," he declared in his summation. "It had never left his mind, and it was the main defense in this court room: 'I had a right to protect my property.' "

And on the witness stand, Isaac Harris had presented this astounding picture of the firm's tireless battle to protect its property:

"We once locked up about six girls and we found in their room, the room of one girl, about two dozen waists, one girl about three dozen waists, and one girl had—er—in every house we found so many waists and we had detectives that went around there and we searched in every house and found from two dozen to three dozen waists that these girls had taken and there were six girls that we locked up in one night about three years ago.

"About eight months before the fire, one girl took two waists in her rat and it stuck out a little end of the string and one girl coming along behind that girl saw this little white string in her hair and said, 'look here what is sticking out here,' and when she started to pull, she pulled out the waists and they stopped that girl and they took that girl into the office and they took out the two waists and they asked her, 'what do you want to do it for?'

"She said her mistress asked her to bring her two waists. So, of course, we didn't want to make any trouble and we discharged her. All that we could do was that we could discharge her. We could not do any more because when we arrested a few of them and had them fined, a few of the girls sued us for damages. We found the best way was to discharge them and not to bother with them any more."

Dismayed by this revealing confession by his client, Steuer interjected that the newspapers, he feared, would make capital and headlines out of Harris' words and that as a result the firm would be driven out of business.

But undeterred, Bostwick continued, demanding to know from Harris, "How much in all the instances would you say was the value of all the goods that you found had been taken by these employees? You would say it was not over $25, wouldn't you?"

To this, Harris replied: "No, it would not exceed that much."

17. PHOENIX

Those citizens who afterwards rebuilt it
Upon the ashes.

—CANTO XIII: 148

MARCH 30, 1911. Inspectors of the Building Department halted production at Harris and Blanck's new plant at 7 University Place. The department's notice of violations included the charge that in this nonfireproofed building in which Triangle had rented space to continue production despite the tragedy, the sewing machines had been so arranged as to block access to the fire escape.

December 18, 1911. When Salvatore Maltese had marched round and round the boxes in the morgue, looking and looking, he had found first one daughter and then another. But not his wife Catherine. And in the days after they had emptied and cleaned the morgue and the last remaining charred things had been buried with numbers instead of names, his life became a pursuit of things dimly remembered. And he pawed the trinkets spread out before him by the patient police, the last links with identity, until, in the month of the Triangle acquittal, memory focused and he whispered, "This was hers." They dug up the box in Evergreen Cemetery, they took off the number and replaced it with her name and carried her to where her two daughters were buried and Salvatore Maltese and his two sons watched as she was laid to rest among her own.

 March 21, 1912. Protests against the acquittal were still

being heard when the District Attorney moved for a second trial. Almost a year after the fire, Harris and Blanck were again in court. When Steuer protested that his clients had already been placed in jeopardy on the basis of the evidence which would now be merely repeated, Bostwick answered that different persons had been named in each of the seven manslaughter indictments found against the defendants. This one was for the death of Jake Kline, once tossed out of Triangle and later rehired.

Judge Samuel Seabury, presiding in the Criminal Branch of the Supreme Court, ordered a special jury to be selected at once. This having been done, he then addressed the jurors, informing them that "the court has neither the right nor the power to proceed with the present trial. These men are to be tried for the same offense again and under our constitution and laws, this cannot be done. I charge you, gentlemen of the jury, to find a verdict for the defendants."

The jury found as directed without leaving the box.

This time there were no victims, no survivors, no relatives, no crowds in the court. "We did all we could," said District Attorney Charles S. Whitman. "The law was against us today. I will make no comment as to the first trial."

Four days later, prayers in Hebrew and Latin rose in synagogue and church on the East Side where the old men, at sundown, lit the traditional memorial candles to mark the anniversary of the holocaust.

August 20, 1913. Max Blanck was arraigned in General Sessions Court. At this time, Triangle occupied new quarters on the ninth floor of a building on the corner of Sixteenth Street and Fifth Avenue.

Max Blanck, being a free man and without guilt and fortified with the knowledge of what he could do to protect his property, applied the full lesson of his acquittal, as he saw it. In court, Inspector Walter Dugan of the newly created Bureau of Fire Prevention testified that in his recent inspection of the building, he had found one of the Triangle doors locked with a chain—during working hours and with 150 girls in the shop.

On September 19, Blanck produced the lock in court. He claimed that with this patented chain device it was easy to get out, hard to get into his shop. On the other hand, Inspector Dugan insisted that under conditions of panic the lock would become nonmanipulative and that with more pulling it would only become more firmly locked.

One week later, Chief Justice Russell, considering this Blanck's first offense, levied the minimum fine of $20. He then spoke directly to Blanck—and apologized to him for having had to impose any fine at all. The State Labor Department and the factory inspectors, he said, should devise a lock which would be satisfactory to them and at the same time would protect factory proprietors from theft.

December 1, 1913. Blanck was back in court, paying another $20 fine, for allowing Becky Katzman, a nineteen-year-old lace cutter, to work on Sunday, October 12, in violation of the labor law.

December 23, 1913. Obsessed with the idea of a perfect lock, Max Blanck invited members of the Bureau of Fire Prevention to visit his plant two days before Christmas. The sole purpose of the visit was to have them witness an experiment with a new type of lock.

The new lock, the *Times* reported, "was meant to safeguard employers from the loss of goods through the departure of employees through fire exits instead of by way of elevators." Chief Inspector John J. Kennedy, heading the party of visitors, had difficulty concentrating on the experiment. He found the lock completely unsatisfactory.

The Chief had been distracted by a pile of rubbish 6 feet high in one corner of the shop. He observed litter throughout the factory. He noted that wicker work baskets instead of metal boxes were still being used. Instead of winning approval, Blanck received a stern warning.

March 4, 1914. The National Consumers League discovered that a counterfeit of its label, used to certify work done "under clean and healthful conditions," was being used on shirtwaists traced to Triangle. In August, Justice Leonard A. Giegerich in the Supreme Court of New York County declared that the Triangle label was an illegal imitation and issued an injunction barring its use.

March 11, 1914. Three years after the fire, twenty-three individual suits against the owner of the Asch building were settled at the rate of $75 per life lost.

"The claimants have been tired out," said the *World*. "Their money and their patience have been exhausted. So far as personal guilt is concerned, the men whose methods made everything ready for the tragedy have gone free. So far as financial liability is concerned, the whole affair is in the hands of an insurance company and stricken families are not well equipped to carry on expensive litigations with corporations."

The establishment of the Bureau of Fire Prevention with the passage of the Sullivan-Hoey law in October, 1911, was one of the first local results of the uneasiness that continued to be felt in New York City following the holocaust. The new law also expanded the powers and the duties of the Fire Commissioner and ended much of the division of responsibilities.

But the shock of the acquittal had created a widespread feeling that something more was needed than additional fire escapes. Indeed, Richard B. Morris points out that "it is doubtful whether the social consequences of the Triangle fire would have been as far-reaching had Steuer lost his case."

At the first protest meeting, held at the headquarters of the Women's Trade Union League on the day after the fire, Rabbi Wise had called for the creation of a committee of twenty-five to improve safety in working places—the task the parties and the politicians had so far failed to execute. In one week—by the time the great public protest meeting was held the following Sunday in the Metropolitan Opera House—the committee was established and functioning. Its membership cut across party and class lines and included such notables as Anne Morgan, Mary Dreier, Frances Perkins, George W. Perkins, John A. Kingsbury, Peter Brady, Amos Pinchot, Rabbi Wise, and a number of other clergymen. Its chairman was Henry L. Stimson, who was soon compelled to relinquish that post in order to assume the office of Secretary of War, and who was succeeded as chairman by Henry Morgenthau Sr.

The Committee on Safety, together with the Women's Trade Union League, the National Consumers League, the Fifth Avenue Association, the unions and public-spirited individuals, maintained a steady clamor for remedial legislation. On June 30, 1911, the New York State Legislature created a special Factory Investigating Commission of nine members. Its chairman and vice chairman were two young men starting auspicious careers in American political life: Robert F. Wagner, Sr., and Alfred E. Smith.

Many who had perceived the possibility of tragedy even before it had happened were associated with the Commission, two as members of the Commission itself. They were Samuel Gompers, president of the American Federation of Labor, who had warned the shirtwaist manufacturers at the historic Cooper Union meeting and Mary Dreier, president of the Women's Trade Union League, who had marched on the shirtwaist makers' picket line and had been arrested.

The Commission, in turn, used the services of many experts, and these, too, included a number who had warned against callousness in dealing with industrial dangers. Dr. George M. Price, whose report for the Joint Board of Sanitary Control had spelled out the danger in New York's garment shops only days before the fire, was named the Commission's expert on sanitation; H. F. J. Porter, whose plea to Triangle to institute fire drills had not even received the courtesy of a reply, became its expert on fire problems. Its corps of inspectors included Rose Schneiderman, who had said at the Metropolitan Opera House meeting that pity was not enough, and Frances Perkins, destined to become, twenty years later, the Secretary of Labor as Franklin D. Roosevelt launched the New Deal.

The puny sum of $10,000 was appropriated for the work of the Commission, which, from the start, had set its sights on a much larger target than the fire hazard. In its preliminary report the Commission declared that a "superficial examination revealed conditions in factories and manufacturing establishments that constituted a daily menace to the lives of thousands of working men, women and children. The need for a thorough and extensive investigation into the general conditions of factory life was clearly recognized."

Wagner and Smith went to see Henry Morgenthau soon after the Commission was named. They told him that $10,000 would barely be enough to pay a good attorney for doing the essential legal work of the Commission. When they had finished, Morgenthau said he would get the best possible lawyer for the special task and that he would work without fee.

Within two hours after the interview, arrangements were completed for Abram I. Elkus to serve as chief counsel; Bernard L. Shientag joined as assistant counsel. Both men served without fee; both later became distinguished judges.

The Commission held its first session in New York City on October 14, 1911. It exercised fully its power to compel attendance of witnesses and the production of books and papers and in its first year heard 222 witnesses and inspected 1,836 industrial establishments in various cities in the state.

It heard New York Fire Chief John Kenlon estimate the cost of installing sprinklers in the Asch building at $5,000. Kenlon added that in his opinion not a life would have been lost if there had been sprinklers there on March 25.

The Chief was asked to give the addresses of the several hundred other buildings in which "another Asch building fire is likely to occur." He refused, saying, "I think it unwise to do that," and the Commission let it stand.

It heard the Commissioner of Labor declare that there was no automatic, legally required record of establishments and that only through search and patrol by his men was it possible to "ascertain where factories are located, their kind and nature and the number of employees therein."

It listened to a State Labor Department inspector tell how he started his inspection of the Triangle shop on February 27, 1911, less than a month before the fire, by first going to the company's office and introducing himself.

"You see," Counsel Elkus commented, "what we are trying to point out, Mr. Harmon, is that these conditions which exist in factories are not discovered because the inspectors inform somebody in authority that they are there and the persons in authority know what you are looking for so that they had plenty of time to remedy temporarily any defects that existed while you were there."

The Commission sent its investigators creeping, crawling, climbing, prying into the dark and hidden corners of cellars and shops, factories and tenement houses. Shocking testimony was heard about toilets, dust-filled workshops, dirty factories, unguarded but crippling machinery, working children, working women, night shifts, lead poisoning, diseased workers, lack of ventilation, poor lighting. "Establishments manufacturing foodstuffs were the dirtiest of all. The conditions of the toilets in most of these factories was very bad. The flush was usually found to be inadequate."

The four-year term of the Commission marks the beginning of what is generally recognized as "the golden era in remedial factory legislation," in the state of New York. Its first year of work resulted in the addition of eight new laws to the labor code; it produced twenty-five new laws the following year and three in 1914.

These laws completely overhauled the State Department of Labor and provided for a staff sufficient to carry out the many new duties assigned to it. The Commission's work had been touched off by the fire hazard but, while it dealt effectively with this danger, it also turned to other hazards, and among its new laws in one area alone were those limiting the hours of labor of women and children, abolishing night shifts for them, bringing canneries under the labor law, providing seats with backs for women workers, and prohibiting the employment of women immediately after childbirth.

On the first day of the Commission's public hearings Chief Counsel Elkus outlined a new purpose in American life—one which twenty years later would be at the heart of a program through which the entire nation would seek to lift itself out of a disaster far greater in scope than the Triangle fire.

"A man may be killed by a tenement house as truly as by

a club or gun," Elkus declared at the opening hearing. "A man may be killed by a factory and the unsanitary conditions in it as surely as he may be killed by a fire.

"It is not less true that the slaughter of men and women workers by the slow process of unsanitary and unhealthful conditions is not only immoral and anti-social, but the state is beginning to declare that it is legally indefensible and therefore must, through carefully considered legislation, be made virtually impossible. That their industrial efficiency may be unimpaired is of prime economic importance to the state.

"The so-called unavoidable or unpreventable accidents which, it has been said, were once believed to be the result of the inscrutable decrees of Divine Providence are now seen to be the result in many cases of unscrupulous greed or human improvidence. It is the duty of the state to safeguard the worker not only against the occasional accidents but also the daily incidents of industry; not only against the accidents which are extraordinary but also against the incidents which are the ordinary occurrences of industrial life," Elkus concluded.

Twenty years before she became the Secretary of Labor in the cabinet of President Franklin D. Roosevelt, the post she held until his death, Frances Perkins, on the day of the Triangle fire, was visiting friends "on the other side of the Square."

"It was a fine, bright spring afternoon" she remembers. "We heard the fire engines and rushed into the Square to see what was going on. We saw the smoke pouring out of the building. We got there just as they started to jump. I shall never forget the frozen horror which came over us as we stood with our hands on our throats watching that horrible sight, knowing that there was no help. They came down in twos and threes, jumping together in a kind of desperate hope.

"The life nets were broken. The firemen kept shouting for them not to jump. But they had no choice; the flames were right behind them for by this time the fire was far gone.

"Out of that terrible episode came a self-examination of

stricken conscience in which the people of this state saw for the first time the individual worth and value of each of those 146 people who fell or were burned in that great fire. And we saw, too, the great human value of every individual who was injured in an accident by a machine.

"There was a stricken conscience of public guilt and we all felt that we had been wrong, that something was wrong with that building which we had accepted or the tragedy never would have happened. Moved by this sense of stricken guilt, we banded ourselves together to find a way by law to prevent this kind of disaster.

"And so it was that the Factory Commission that sprang out of the ashes of the tragedy made an investigation that took four years to complete, four years of searching, of public hearings, of legislative formulations, of pressuring through the legislature the greatest battery of bills to prevent disasters and hardships affecting working people, of passing laws the likes of which has never been seen in any four sessions of any state legislature.

"It was the beginning of a new and important drive to bring the humanities to the life of the brothers and sisters we all had in the working groups of these United States. The stirring up of the public conscience and the act of the people in penitence brought about not only these laws which make New York State to this day the best state in relation to factory laws; it was also that stirring of conscience which brought about in 1932 the introduction of a new element into the life of the whole United States.

"We had in the election of Franklin Roosevelt the beginning of what has come to be called a New Deal for the United States. But it was based really upon the experiences that we had had in New York State and upon the sacrifices of those who, we faithfully remember with affection and respect, died in that terrible fire on March 25, 1911. They did not die in vain and we will never forget them."

18. FIRE

Truly I wept . . .
—CANTO XX:25

ON MARCH 19, 1958, in the late afternoon of a miserable rainy day, I was checking out proofs of a special feature marking the forty-seventh anniversary of the Triangle fire in the pages of *Justice*, the ILGWU publication.

A call came in from the city desk of a local newspaper. The voice at the other end asked how many had died in the Triangle fire. I answered 146 and asked why he wanted to know.

"They're coming out of the windows again!" was what I heard. He told me where.

I rushed to Houston Street and Broadway.

I stood there in the rain and, as if in a nightmare, watched a re-enactment of the Triangle horror, which had occurred not more than five city blocks away.

This was no new building such as the Asch had been at the time of its fire. 623 Broadway was already thirty years old at the time of the Triangle fire. Now, it was seventy-seven years old. It was six floors high, had no sprinklers, its fire escape was worthless, there had been no fire drills. It was shaped like a 90-foot-long wind tunnel, and in the middle of the floor areas were glass blocks, overlaid with wood in some forgotten and unrecorded time when the structure was converted from a warehouse to a factory building.

The Triangle Fire

The fire had started on the third floor in a textile finishing firm using an oven box which blew up. It burned for seven minutes before an alarm was turned in—from the street. Almost until that moment, the workers on the fourth floor, busy making women's undergarments in a clean, well-run, union shop, were unaware of the hell burning beneath them.

Then the smoke came. A courageous boss cautioned against panic, stayed with his workers, died with them.

For soon the glass blocks worked loose and the entire middle section of the fourth floor collapsed—flat, like a falling pancake. Twenty-four died.

My friend Josephine Nicolosi still lives in the neighborhood she called home in that distant time when she almost lost her life on the eighth floor of the Triangle shop. The rear windows of her top-floor tenement apartment on Second Avenue just above Houston Street face west. In the late afternoon of March 19 she had seen a pillar of black smoke rising in the sky. She grabbed her coat and hurried to the street.

Transfixed by the sight of the bodies being lowered in the baskets, we almost missed each other, and when I saw her suddenly, her face was wet with rain and tears. She gripped me by the wrists and shaking me demanded with anger and despair:

"What good have been all the years? The fire still burns."

POSTSCRIPT

MY CHIEF SOURCES of help in gathering the material for this book have been Steward Liddell, the newspaper reporters of New York City in 1911, and a group of survivors and witnesses of the Triangle fire.

Throughout the trial of Harris and Blanck, in December, 1911, Steward Liddell, court stenographer, kept the record straight. It is from that record that many who were trapped in the Asch building but somehow managed to escape speak directly in the pages of this book with sworn veracity.

In a time of personalized but anonymous journalism, the reporters of that day filled many columns about the fire with facts encased in unashamed emotion. One among them merits special mention.

Purely by chance, William Gunn Shepherd of the United Press was at the scene of the fire from its start. From behind a store front window across the street from the Asch building, he watched its progress and described the horror over the telephone to his city editor, Roy W. Howard, later head of the Scripps-Howard newspaper chain, who received it in the old World building on Park Row.

Until relief arrived, Shepherd stayed on the telephone, told the story as he saw it happen, his voice cracking time and again. In the newsroom, deserted on Saturday afternoon except for the four men working the telegraph trunk lines to the country, Howard cleared the lines with a stand-by order.

Shepherd returned to the newsroom worn and tired as were also the four men who had sweated it out at the receiving end of the phone. Howard broke the silence with the finest compliment he could give—a request to Shepherd to get to his desk to do a by-lined rewrite of the story.

Sometimes with quotation marks and at times without them, I have retained the words that first appeared in the New York *American,* the New York *Call,* the New York *Herald,* the New York *Evening Journal,* the New York *Sun,* the New York *Times,* the New York *Tribune,* the New York *World* and the *Jewish Daily Forward.* I have also drawn material from these magazines: the *American Federationist, Collier's, Independent, Justice, Life and Labor, McClure's, Outlook,* and *Survey.*

But my most treasured collaborators have been the survivors and witnesses of the fire, who have been kind enough to open their hearts to me. They had the right to refuse to talk about their painful memories. Some had lost dear ones in the fire. Not one of them, in almost fifty years, had ever returned to the scene of the tragedy.

That was true of the aged couple—Ercole Montanare and his wife—who came to my office. He had come with her to tell me how on the day of the fire, he and a friend, leaving their shop on Fourth Avenue, had decided to walk in the welcome sunshine to Canal Street only to be intercepted by the horror of the falling bodies.

On the way to my office, they had stopped to see if they could locate the Asch building. They weren't certain, and at the corner of Greene Street and Washington Place they went into stores and asked storekeepers which had been the Triangle building. No one knew.

Back on the sidewalk, they moved about slowly, two fragile old people looking up. Others looked at them and then lifted their own gaze upward, seeing nothing but the tall buildings. Then Montanare recognized it. "I knew it was the right building," he told me, "because I could see again the bodies falling, I could hear again their screaming."

The tragic day returned in full vividness. "That night, many years ago, I couldn't sleep. The burnt and bleeding bodies kept falling. In the morning I took the elevated at Chatham Square and rode on the open platform to Bronx Park. I thought my chest would burst. The whole day I ran along the river in that park like a wild animal, spitting and screaming and swallowing big mouthfuls of the clean, sweet air."

Some had not spoken of the fire in almost fifty years. Two refused to speak of it even now. A frequent pattern of the interviews was an initial reluctance to speak, then a sudden emotional outpouring of memories subsiding into a lengthy, calm series of recollections.

There were also humor and strange twists of fate in these visits. In Jersey City, one evening, I sat drinking tea with an elderly couple. Then Rose Cohen Indursky told me of the little girl she remembered who had come running up to the roof of the burning building wrapped in fabric caught by the flames. She had helped the youngster beat out the flames.

Three weeks earlier, in Brooklyn, Ida Nelson Kornweiser had told me she had never been able to learn the name of the girl who had saved her life by slapping out the flames eating the fabric she had wrapped herself in to run to the roof. There were tears in Rose Indursky's eyes when I promised to arrange a reunion.

There were tears also one night in Hackensack, New Jersey, when bitter memory pierced the dignity of lovely Anna Gullo Pidone. At the start of our interview, she brought into the room and placed on the table before us, where it remained throughout the evening, a tintype of her sister in her confirmation dress—the sister who had died in the fire.

But there was also another evening filled with joy and excitement when my wife and I visited Ethel Monick Feigen and found her surrounded by a dozen children and grandchildren.

Soon after our arrival she took me aside and whispered to me that from the time of her first-born she had made it a practice to tell each of them, when they reached the age of twelve, about the big fire in which she had almost lost her life. Then she added, "They have all listened to me politely. But, you know, I think not one of them has ever really believed me. Now, you please tell them." I did.

I looked a long time for Joseph P. Meehan, the heroic mounted policeman I first met in the newspaper accounts. Five days before the start of the demolition of the 8th Precinct Police Station on Mercer Street I asked the officer in charge for the day book for March 25, 1911.

He came around from behind the high desk and I followed him down into a deep subbasement with one bare bulb hanging from the center of the ceiling. The walls were lined with precinct record books starting at a pre-Civil War date. We found the book for March, 1911, and excitedly I turned the pages, running my finger down the timed entries until I found him again.

He had come into the station house at 6:30 that fateful Saturday and had wearily reported that in putting a victim into an automobile he had ripped the tail of the coat of his uniform.

I located him last year living in Manhattan's mid-Sixties. Forty-nine years after his great gallop up Washington Place I came to him to talk about it. Every wall of his spacious apartment was covered with pictures, photographs, paintings, drawings—of horses.

He had been on horses for more than four decades, retiring as a high-ranking officer of the mounted police. Yes, he remembered the name of the horse he rode that day. "It was Yale, and I was the only one who could manage him." No, he had never been compensated for his damaged uniform: "the entire skirt had been ripped off."

Time ran out for some interviews. I would have talked, if I could, with Edward F. Croker. On December 10, 1929, ten persons lost their lives in the Pathé Studio fire at Park Avenue and 134th Street, where there was more film than the law allowed, where there were fewer fire extinguishers than the law required, and no sprinklers because Croker's fire prevention company, on an annual $500 retainer, had advised appeal against an order to install them.

POSTSCRIPT

These have been the friends who have helped tell this story: Joseph Brenman, Sarah Friedman Dworetz, Ethel Monick Feigen, Joseph Flecher, Celia Walker Friedman, Sigmund Gelbart, Abe Gordon, Joseph Granick, Rose Glantz Hauser, Max Hochfield, Rose Cohen Indursky, Elias Kanter, Sylvia Riegler Kimmeldorf, Ida Nelson Kornweiser, Kate Weiner Lubliner, Dora Maisler, James P. Meehan, Josephine Nicolosi, Anna Gullo Pidone, Celia Saltz Pollack, Frank Rubino, Dora Appel Skalka, Sarah Cammerstein Stern, Joseph Wexler, Isidore Wegodner. Others were Mary Abrams, Mrs. Charles F. Bostwick, Jr., Daniel Charnin, Ercole Montanare, and Elmer E. Wigg.

I am indebted to YIVO (Yiddish Scientific Institute) for the use of Bernard Baum's excellent memoir on preconflagration life at Triangle that has now been published by him in Yiddish.

Sigmund Arywitz, California State Commissioner of Labor, taped interviews with Triangle survivors in that state; by inviting me to broadcast on his radio program in the days following the Monarch fire, Tex McCrary made it possible for me to locate a number of the survivors of the earlier tragedy; New York Fire Commissioner Edward F. Cavanagh, Jr., and other officers of the New York Fire Department have helped educate me on technical matters pertaining to industrial fires. I have also used the "Preliminary Report of the Factory Investigation Commission, Albany, N.Y., 1912" and the "Report on Relief of Triangle Fire Victims, Special Committee, American Red Cross, New York, 1912."

I have been a member of the International Ladies' Garment Workers' Union since 1928; until I joined the staff of its publication, *Justice,* in 1942, I worked as a garment cutter. In these last twenty years, close and constant association with ILGWU President David Dubinsky has been a source of unending insight and inspiration. My wife has patiently shared the throes of composition, aiding with her vetoes and applause. My colleague, Meyer Miller, was helpful with a critical reading of the first draft.

I owe a special debt to three great ladies who helped turn the Triangle tragedy into a force for change and reform. Eleanor Roosevelt, Frances Perkins, and Rose Schneiderman graced with their presence the Fiftieth Anniversary Memorial Meeting of the Triangle Fire which the ILGWU, together with New York University and the New York Fire Department, sponsored on Saturday, March 25, 1961. Fourteen survivors shared the platform with them.

The irony of history was evident in the fact that on that day there lay on the desk of Governor Nelson Rockefeller, awaiting his signature, a bill that would have in effect rescinded long overdue safety measures enacted following the Monarch fire. To some on the platform and to the old-timers in the audience gathered at the foot of what was once the Asch building, it must have seemed, indeed, that the passing years had made no difference.

David Dubinsky charged that if signed the bill would restore conditions as they were at the time of the Triangle fire, and Eleanor Roosevelt, Frances Perkins, and Rose Schneiderman helped picture the human meaning of such a reversal. The huge fire bell tolled for those who had died, a special corps of uniformed firemen along with Commissioner Cavanagh gave the salute, there were tears.

After the meeting I introduced the survivors to the three special guests, to Esther Peterson, now Assistant Secretary of Labor, and to Dubinsky. I saved three for the last.

They had met before but never formally. On the platform in the shadow of the Asch building I introduced to each other Sarah Cammerstein Stern, Sarah Friedman Dworetz, and Celia Walker Friedman, who fifty years earlier, almost to the hour, had lain bleeding and battered at the bottom of the elevator shaft down which each had leaped from the open ninth-floor elevator door.

INDEX

Index

Index